LINEAR INTEGRAL EQUATIONS

BY

WILLIAM VERNON LOVITT, PH.D.

Professor of Mathematics, Colorado College

FIRST EDITION

DOVER PUBLICATIONS INC.
NEW YORK

This Dover edition, first published in 1950, is an unabridged and unaltered republication of the work originally published by the McGraw-Hill Book Co., Inc., in 1924.

Library of Congress Catalog Card Number: A51-10414

Manufactured in the United States of America
Dover Publications, Inc.
180 Varick Street
New York, N.Y. 10014

PREFACE

For many years the subject of functional equations has held a prominent place in the attention of mathematicians. In more recent years this attention has been directed to a particular kind of functional equation, an integral equation, wherein the unknown function occurs under the integral sign. The study of this kind of equation is sometimes referred to as the inversion of a definite integral.

In the present volume I have tried to present in a readable and systematic manner the general theory of linear integral equations with some of its applications. The applications given are to differential equations, calculus of variations, and some problems in mathematical physics. The applications to mathematical physics herein given are to Neumann's problem, Dirichlet's problem, and certain vibration problems which lead to differential equations with boundary conditions. The attempt has been made to present the subject matter in such a way as to make the volume available as a text on this subject in Colleges and Universities.

The reader who so desires can omit the chapters on the applications. The remaining chapters on the general theory are an entity in themselves.

The discussion has been confined to those equations which are linear and in which a single integration occurs. The limits of the present volume forbid any adequate treatment of integral equations in several independent variables; systems of integral equations; integral equations of higher order; integro-differential equations; singular integral equations; integral equations with special or discontinuous kernels.

I desire here to express my thanks to Prof. Oscar Bolza (now of Freiburg University, formerly of the University of Chicago) for his permission to make use of my notes on his lectures on integral equations delivered during the summer of 1913 at the University of Chicago. A bound volume of these notes in my handwriting has resided these past ten years on the shelves of the University of Chicago mathematical library and has been available to many students during this time. A number of copies of these notes are in circulation in this country at present.

The following books have been available and have been found to be of value in the preparation of this volume:

M. Bocher: An Introduction to the Study of Integral Equations. No. 10, Cambridge Tracts, 1909. University Press. Cambridge.

É. Goursat: Cours D'Analyse Mathématique. Tome III. Chaps. 30, 31, 32, 33. Paris, Gauthier-Villars, 1923.

Heywood-Fréchet: L'Équation de Fredholm et ses applications à la Physique Mathématique. Paris, Hermann et Fils, 1912.

Kneser: Die Integralgleichungen und ihre Anwendungen in der Math. Physik. Braunschweig. Vieweg et Sohn, 1922.

G. Kowalewski: Einführung in die Determinanten Theorie. 18 and 19 cap. (p. 455–505). Verlag von Veit et C., 1909.

Lalesco: Introduction à la Théorie des Équations Intégrales. Paris. Hermann et Fils, 1912.

Volterra: Leçons sur les Équations Intégrales et les Équations Intégro-Differentielles. Paris, Gautheir-Villars, 1913.

For those who desire a bibliography on this subject we refer the reader to a short bibliography in the work by Heywood-Fréchet and to a more extensive bibliography in the work by Lalesco.

W. V. Lovitt.

Colorado Springs, Colo., *June*, 1924.

CONTENTS

CHAPTER I

INTRODUCTORY

CHAPTER II

SOLUTION OF INTEGRAL EQUATION OF SECOND KIND BY SUCCESSIVE SUBSTITUTIONS

CHAPTER III

SOLUTION OF FREDHOLM'S EQUATION EXPRESSED AS RATIO OF TWO INTEGRAL SERIES IN λ

CHAPTER IV

Applications of the Fredholm Theory

I. Free Vibrations of an Elastic String

II. Constrained Vibrations of an Elastic String

III. Auxiliary Theorems on Harmonic Functions

IV. Logarithmic Potential of a Double Layer

CHAPTER I

INTRODUCTORY

1. Linear Integral Equation of the First Kind.—An equation of the form

$$\int_a^b K(x, t)u(t)dt = f(x) \tag{1}$$

is said to be a **linear integral equation of the first kind.** The functions $K(x, t)$ and $f(x)$ and the limits a and b are known. It is proposed so to determine the unknown function u that (1) is satisfied for all values of x in the closed interval $a \leqq x \leqq b$. $K(x, t)$ is called the *kernel* of this equation.

Instead of equation (1), we have often to deal with equations of exactly the same form in which the upper limit of integration is the variable x. Such an equation is seen to be a special case of (1) in which the kernel $K(x, t)$ vanishes when $t > x$, since it then makes no difference whether x or b is used as the upper limit of integration.

The characteristic feature of this equation is that the unknown function u occurs under a definite integral. Hence equation (1) is called an *integral* equation and, since u occurs linearly, equation (1) is called a *linear* integral equation.

2. Abel's Problem.—As an illustration of the way in which integral equations arise, we give here a statement of Abel's problem.

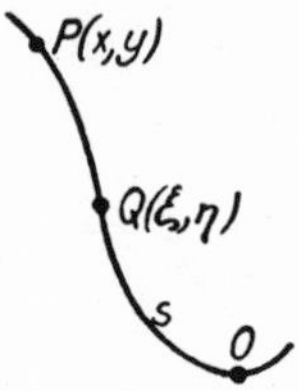

FIG. 1.

Given a smooth curve situated in a vertical plane. A particle starts from rest at any point P. Let us find, under the action of gravity, the time T of descent to the lowest point O. Choose O as the origin of coordinates, the x-axis vertically upward, and the y-axis horizontal. Let the coordinates of P be (x, y), of Q be (ξ, η), and s the arc OQ.

The velocity of the particle at Q is

$$\frac{ds}{dt} = -\sqrt{2g(x - \xi)}.$$

Hence

$$t = -\int_P^Q \frac{ds}{\sqrt{2g(x - \xi)}}.$$

The whole time of descent is, then,

$$T = \int_o^P \frac{ds}{\sqrt{2g(x - \xi)}}.$$

If the shape of the curve is given, then s can be expressed in terms of ξ and hence ds can be expressed in terms of ξ. Let

$$ds = u(\xi)d\xi.$$

Then

$$T = \int_o^x \frac{u(\xi)d\xi}{\sqrt{2g(x - \xi)}}.$$

Abel set himself the problem[1] of finding that curve for which the time T of descent is a given function of x, say $f(x)$.

[1] For a solution of this problem, see BÔCHER, "Integral Equations," p. 8, Cambridge University Press, 1909.

Our problem, then, is to find the unknown function u from the equation

$$f(x) = \int_o^x \frac{1}{\sqrt{2g(x-\xi)}} u(\xi)d\xi.$$

This is a linear integral equation of the first kind for the determination of u.

3. Linear Integral Equation of the Second Kind.—An equation of the form

$$u(x) = f(x) + \int_a^b K(x, t)u(t)dt \tag{2}$$

is said to be a **linear integral equation of the second kind.**

$K(x, t)$ is called the kernel of this equation. The functions $K(x, t)$ and $f(x)$ and the limits a and b are known. The function u is unknown.

The equation

$$u(x) = f(x) + \int_a^x K(x, t)u(t)dt$$

is known as Volterra's linear integral equation of the second kind.

If $f(x) \equiv 0$, then

$$u(x) = \int_a^b K(x, t)u(t)dt.$$

This equation is said to be a homogeneous linear integral equation of the second kind.

Sometimes, in order to facilitate the discussion, a parameter λ is introduced, thus

$$u(x) = f(x) + \lambda \int_a^b K(x, t)u(t)dt.$$

This equation is said to be a linear integral equation of the second kind with a parameter.

Linear integral equations of the first and second kinds are special cases of the linear integral equation of the third kind:

$$\Psi(x)u(x) = f(x) + \int_a^b K(x, t)u(t)dt.$$

Equation (1) is obtained if $\Psi(x) \equiv 0$.
Equation (2) is obtained if $\Psi(x) \equiv 1$.

4. Relation between Linear Differential Equations and Volterra's Integral Equation.—Consider the equation

$$\frac{d^ny}{dx^n} + a_1(x)\frac{d^{n-1}y}{dx^{n-1}} + \dots + a_n(x)y = \varphi(x), \tag{3}$$

where the origin is a regular point for the $a_i(x)$.

Let us make the transformation

$$\frac{d^ny}{dx^n} = u(x).$$

Then

$$(4)\quad \begin{cases} \dfrac{d^{n-1}y}{dx^{n-1}} = \displaystyle\int_o^x u(x)dx + C_1 \\ \dots\dots\dots\dots\dots\dots \\ y = \displaystyle\int_o^x u(x)dx^n + C_1\frac{x^{n-1}}{\underline{|n-1}} + C_2\frac{x^{n-2}}{\underline{|n-2}} + \dots + C_n, \end{cases}$$

where $\int_o^x u(x)dx^n$ stands for a multiple integral of order n.

Equations (4) transform (3) into

$$(5)\quad u(x) + a_1(x)\int_o^x u(x)dx + \dots + a_n(x)\int_o^x u(x)dx^n = \varphi(x) + \sum_1^n {}^i\, C_i\alpha_i(x),$$

where

$$\alpha_i(x) = a_i(x) + \frac{x}{1}a_{i+1}(x) + \dots + a_n(x)\frac{x^{n-i}}{\underline{|n-i}}.$$

If we now put

$$\varphi(x) + \sum_1^n {}^i C_i\alpha_i(x) = f(x),$$

and make use of the well-known formula

$$\int_0^x u(t)dt^n = \int_0^x \frac{(x-t)^{n-1}}{|\underline{n-1}} u(t)dt,$$

equation (5) becomes

$$u(x) + \int_0^x \Big[a_1(x) + a_2(x)(x-t) + \cdots + a_n(x)\frac{(x-t)^{n-1}}{|\underline{n-1}}\Big] u(t)dt = f(x),$$

which is a Volterra integral equation of the second kind.

In order that the right-hand member of (5) have a definite value it is necessary that the coefficients C_i have definite values. Then, inversely, the solution of the Volterra's equation (5) is equivalent to the solution of Cauchy's problem for the linear differential equation (3). The uniqueness of the solution of Volterra's equation follows from the fact that Cauchy's problem admits for a regular point one and only one solution.[1]

5. Non-linear Equations.—This work will be confined to a discussion of linear integral equations. It is desirable, however, at this point to call the reader's attention to some integral equations which are non-linear.

The unknown function may appear in the equation to a power n greater than 1, for example;

$$u(x) = f(x) + \lambda \int_a^b K(x,t)u^n(t)dt.$$

The unknown function may appear in a more general way, as indicated by the following equation:

$$u(x) = f(x) + \lambda \int_a^b \varphi[x, t, u(t)]dt.$$

[1] For further discussion consult LALESCO, T., "Théorie Des Équations Intégrales," pp. 12ff, Herman and Fils, Paris, 1912.

In particular, the differential equation

$$\frac{du}{dx} = \varphi(x, u)$$

can be put in the integral form

$$u(x) = C + \int_0^x \varphi[t, u(t)]dt.$$

Still other general types of non-linear integral equations have been considered. Studies have also been made of systems of integral equations both linear and non-linear. Some study has been made of integral equations in more than one variable, for example;

$$u(x, y) = f(x, y) + \lambda \int_a^b \int_c^d K(x, y; t_1, t_2)u(t_1, t_2)dt_1dt_2.$$

6. Singular Equations.—An integral equation is said to be singular when either one or both of the limits of integration become infinite, for example;

$$u(x) = f(x) + \lambda \int_0^\infty \sin (xt)u(t)dt.$$

An integral equation is also said to be singular if the kernel becomes infinite for one or more points of the interval under discussion, for example;

$$f(x) = \int_0^x \frac{H(x, t)}{(x - t)^\alpha} u(t)dt \qquad (0<\alpha<1).$$

Abel's problem, as stated in §2, is of this character. Abel set himself the problem of solving the more general equation

$$f(x) = \int_a^x \frac{u(t)dt}{(x - t)^\alpha} \qquad (0<\alpha<1).$$

7. Types of Solutions.—By the use of distinct methods, the solution of a linear integral equation of the second kind with a parameter λ has been obtained in three different forms:

1. The first method, that of successive substitutions, due to Neumann, Liouville, and Volterra, gives us $u(x)$ as an integral series in λ, the coefficients of the various powers of λ being functions of x. The series converges for values of λ less in absolute value than a certain fixed number.

2. The second method, due to Fredholm, gives $u(x)$ as the ratio of two integral series in λ. Each series has an *infinite* radius of convergence. In the numerator the coefficients of the various powers of λ are functions of x. The denominator is independent of x. For those values of λ for which the denominator vanishes, there is, in general, no solution, but the method gives the solution in those exceptional cases in which a solution does exist. The solution is obtained by regarding the integral equation as the limiting form of a system of n linear algebraic equations in n variables as n becomes infinite.

3. The third method, developed by Hilbert and Schmidt, gives $u(x)$ in terms of a set of *fundamental functions.* The functions are, in the ordinary case, the solutions of the corresponding homogeneous equation

$$u(x) = \lambda \int_a^b K(x, t)u(t)dt.$$

In general, this equation has but one solution:

$$u(x) \equiv 0.$$

But there exists a set of numbers,

$$\lambda_1, \lambda_2, \ldots, \lambda_n, \ldots;$$

called *characteristic constants* or *fundamental numbers,* for each of which this equation has a finite solution:

$$u_1(x), u_2(x), \ldots, u_n(x), \ldots$$

These are the fundamental functions. The solution then is obtained in the form

$$u(x) = \sum C_n u_n(x),$$

where the C_n are arbitrary constants.

EXERCISES

Form the integral equations corresponding to the following differential equations with the given initial conditions:

1. $\frac{d^2y}{dx^2} + y = 0,\ x = 0,\ y = 0,\ y' = 1,\ y'' = 0.$

$$Ans.\ u(x) = x + \int_0^x (t - x)u(t)dt.$$

2. $\frac{d^2y}{dx^2} - 5\frac{dy}{dx} + 6y = 0,\ x = y = 0,\ y' = -1,\ y'' = -5.$

$$Ans.\ u(x) = 29 + 6x + \int_0^x (6x - 6t + 5)u(t)dt.$$

3. $\frac{d^2y}{dx^2} + y = \cos x,\ x = y = 0,\ y' = 1,\ y'' = 2.$

$$Ans.\ u(x) = \cos x - x - 2 + \int_0^x (t - x)u(t)dt.$$

4. $\frac{dy}{dx} - y = 0,\ x = 0,\ y = y' = 1.$

$$Ans.\ u(x) = 1 + \int_0^x u(t)dt.$$

5. $\frac{d^3y}{dx^3} - 3\frac{d^2y}{dx^2} - 6\frac{dy}{dx} + 5y = 0.$

$$Ans.\ u(x) = 1 - 2x - 4x^2 + \int_0^x [3 + 6(x - t) - 4(x - t)^2]u(t)dt.$$

CHAPTER II

SOLUTION OF INTEGRAL EQUATION OF SECOND KIND BY SUCCESSIVE SUBSTITUTIONS

8. Solution by Successive Substitutions.—We proceed now to a solution of the linear integral equation of the second kind with a parameter. We take up first the case where both limits of integration are fixed (*Fredholm's equation*). We assume that

(1) *a*) $u(x) = f(x) + \lambda \int_a^b K(x,t)u(t)dt$, ($a$, b, constants).

b) $K(x, t) \not\equiv 0$, is real and continuous in the rectangle R, for which $a \leqq x \leqq b$ and $a \leqq t \leqq b$.

c) $f(x) \not\equiv 0$, is real and continuous in the interval I, for which $a \leqq x \leqq b$.

d) λ, constant.

We see at once that if there exists a continuous solution $u(x)$ of (1) and $K(x, t)$ is continuous, then $f(x)$ must be continuous. Hence the inclusion of condition (c) above.

Substitute in the second member of (1), in place of $u(t)$, its value as given by the equation itself. We find

$$\begin{aligned} u(x) &= f(x) + \lambda \int_a^b K(x, t)\left[f(t) + \lambda \int_a^b K(t, t_1)u(t_1)dt_1\right]dt \\ &= f(x) + \lambda \int_a^b K(x, t)f(t)dt \\ &\quad + \lambda^2 \int_a^b K(x, t) \int_a^b K(t, t_1)u(t_1)dt_1dt. \end{aligned}$$

Here again we substitute for $u(t_1)$ its value as given by (1).

We get

$$u(x) = f(x) + \lambda \int_a^b K(x, t)f(t)dt$$
$$+ \lambda^2 \int_a^b K(x, t) \int_a^b K(t, t_1)\left[f(t_1) + \lambda \int_a^b K(t_1, t_2)u(t_2)dt_2\right]dt_1dt$$
$$= f(x) + \lambda \int_a^b K(x, t)f(t)dt$$
$$+ \lambda^2 \int_a^b K(x, t) \int_a^b K(t, t_1)f(t_1)dt_1dt$$
$$+ \lambda^3 \int_a^b K(x, t) \int_a^b K(t, t_1) \int_a^b K(t_1, t_2)u(t_2)dt_2dt_1dt.$$

Proceeding in this way we obtain

$$\text{(2)} \quad u(x) = f(x) + \lambda \int_a^b K(x, t)f(t)dt$$
$$+ \lambda^2 \int_a^b K(x, t) \int_a^b K(t, t_1) f(t_1)dt_1dt + \cdots$$
$$+ \lambda^n \int_a^b K(x, t) \int_a^b K(t, t_1) \cdots$$
$$\int_a^b K(t_{n-2}, t_{n-1})f(t_{n-1})dt_{n-1} \cdots dt_1dt + R_{n+1}(x),$$

where

$$R_{n+1}(x) = \lambda^{n+1} \int_a^b K(x, t) \int_a^b K(t, t_1) \cdots \int_a^b K(t_{n-1}, t_n)$$
$$u(t_n)dt_n \cdots dt_1\, dt.$$

This leads us to the consideration of the following infinite series:

$$\text{(3)} \quad f(x) + \lambda \int^b K(x, t)f(t)dt$$
$$+ \lambda^2 \int_a^b K(x, t) \int_a^b K(t, t_1)f(t_1)dt_1dt + \cdots$$

Under our hypotheses b) and c), each term of this series is continuous in I. This series then represents a continuous function in I, provided it converges uniformly in I.

Since $K(x, t)$ and $f(x)$ are continuous in R and I respectively, $|K|$ has a maximum value M in R and $|f(x)|$ has a maximum value N in I:

$$|K(x, t)| \leqq M \quad \text{in } R$$
$$|f(x)| \leqq N \quad \text{in } I.$$

Put $$S_n(x) = \lambda^n \int_a^b K(x, t) \int_a^b K(t, t_1) \ . \ . \ . \int_a^b K(t_{n-2}, t_{n-1}) f(t_{n-1})dt_{n-1} \ . \ . \ . \ dt_1 dt.$$

Then $$|S_n(x)| \leqq |\lambda^n| N M^n (b - a)^n.$$

The series of which this is a general term converges only when

$$|\lambda| M(b - a) < 1.$$

Thus we see that the series (3) converges absolutely and uniformly when

$$|\lambda| < \frac{1}{M(b - a)}.$$

If (1) has a continuous solution, it must be expressed by (2). If $u(x)$ is continuous in I, its absolute value has a maximum value U. Then

$$|R_{n+1}(x)| < |\lambda^{n+1}| U M^{n+1}(b - a)^{n+1}.$$

If $$|\lambda| M(b - a) < 1, \text{ then}$$
$$\lim_{n \to \infty} R_{n+1}(x) = 0.$$

Thus we see that the function $u(x)$ satisfying (2) is the continuous function given by the series (3).

We can verify by direct substitution that the function $u(x)$ defined by (3) satisfies (1) or, what amounts to the same thing, place the series given by (3) equal to $u(x)$, multiply both sides by $\lambda K(x, t)$ and integrate term by term,[1] as we have a right to do. We obtain

$$\lambda \int_a^b K(x, t)u(t)dt = \lambda \int_a^b K(x,t)\left[f(t) + \lambda \int_a^b K(t, t_1)f(t_1)dt_1 + \cdots \right]dt$$

$$= \lambda \int_a^b K(x, t)f(t)dt$$

$$+ \lambda^2 \int_a^b K(x, t) \int_a^b K(t, t_1)f(t_1)dt_1dt$$

$$+ \cdots$$

$$= u(x) - f(x).$$

Thus we obtain the following:

Theorem I.—If

a) $u(x) = f(x) + \lambda \int_a^b K(x, t)u(t)dt$ (a, b, constants).

b) $K(x, t)$ is real and continuous in a rectangle R, for which $a \leqq x \leqq b$, $a \leqq t \leqq b$.
$|K(x, t)| \leqq M$ in R, $K(x, t) \not\equiv 0$.

c) $f(x) \not\equiv 0$, is real and continuous in I: $a \leqq x \leqq b$.

d) λ constant, $|\lambda| < \dfrac{1}{M(b - a)}$,

then the equation (1) has one and only one continuous solution in I and this solution is given by the absolutely and uniformly convergent series (3).

The equation

$$(4) \qquad u(x) = f(x) + \int_a^b K(x, t)u(t)dt$$

is a special case of the equation (1), for which $\lambda = 1$. The discussion just made holds without change after putting $\lambda = 1$.

[1] GOURSAT-HEDRICK, "Mathematical Analysis, vol. 1, §174, Ginn & Co.

Equations (1) and (4) may have a continuous solution, even though the hypothesis *d*),

$$|\lambda|M(b - a) < 1,$$

is not fulfilled. The truth of this statement is shown by the following example:

$$u(x) = \frac{x}{2} - \frac{1}{3} + \int_0^1 (x + t)u(t)dt,$$

which has the continuous solution $u(x) = x$, while

$$|\lambda|M(b - a) = 2 \not< 1.$$

9. Volterra's Equation.—The equation

$$u(x) = f(x) + \lambda \int_a^x K(x, t)u(t)dt \tag{5}$$

is known as *Volterra's equation.*

Let us substitute successively for $u(t)$ its value as given by (5). We find

$$\begin{aligned}(6)\quad u(x) = f(x) &+ \lambda \int_a^x K(x, t)f(t)dt \\ &+ \lambda^2 \int_a^x K(x, t) \int_a^t K(t, t_1)f(t_1)dt_1\, dt + \cdots \\ &+ \lambda^n \int_a^x K(x, t) \int_a^t K(t, t_1) \cdots \\ &\int_a^{t_{n-2}} K(t_{n-2}, t_{n-1})f(t_{n-1})dt_{n-1} \cdots dt_1\, dt + R_{n+1}(x),\end{aligned}$$

where

$$\begin{aligned}R_{n+1}(x) = \lambda^{n+1} &\int_a^x K(x, t) \int_a^t K(t, t_1) \cdots \\ &\int_a^{t_{n-1}} K(t_{n-1}, t_n)\, u(t_n)dt_n \cdots dt_1\, dt.\end{aligned}$$

We consider the infinite series.

$$\begin{aligned}(6')\quad u(x) = f(x) &+ \lambda \int_a^x K(x, t)f(t)dt \\ &+ \lambda^2 \int_a^x K(x, t) \int_a^t K(t, t_1)f(t_1)dt_1\, dt + \cdots\end{aligned}$$

The general term $V_n(x)$ of this series may be written

$$V_n(x) = \lambda^n \int_a^x K(x, t) \int_a^t K(t, t_1) \; . \; . \; . \int_a^{t_{n-2}} K(t_{n-2}, t_{n-1}) f(t_{n-1}) dt_{n-1} \; . \; . \; . \; dt_1\, dt.$$

Then, since $|K(x, t)| \leqq M$ in R and $|f(t)| \leqq N$ in I, we have

$$|V_n(x)| \leqq |\lambda^n| N M^n \frac{(x - a)^n}{n!} \leqq |\lambda^n| N \frac{[M(b - a)]^n}{n!}, (a \leqq x \leqq b).$$

The series, for which the positive constant $|\lambda^n| N \dfrac{[M(b - a)]^n}{n!}$ is the general expression for the nth term, is convergent for all values of λ, N, M, $(b - a)$. Hence the series (6′) is absolutely and uniformly convergent.

If (5) has a continuous solution, it must be expressed by (6′). If $u(x)$ is continuous in I, its absolute value has a maximum value U. Then

$$|R_{n+1}(x)| \leqq |\lambda^{n+1}| U M^{n+1} \frac{(x - a)^{n+1}}{\underline{|n + 1}} \leqq |\lambda^{n+1}| U \frac{[M(b - a)]^{n+1}}{\underline{|n + 1}}, \quad (a \leqq x \leqq b).$$

Whence

$$\lim_{n \to \infty} R_{n+1}(x) = 0.$$

Thus we see that the function $u(x)$. satisfying (6), is the continuous function given by the series (6′). As before, we can show that the expression for $u(x)$ given by (6′) satisfies (5). Hence we have the following:

Theorem II.—If

(5) *a)* $u(x) = f(x) + \lambda \int_a^x K(x, t) u(t) dt$ (a, constant).

b) $K(x, t)$ is real and continuous in the rectangle R, for which $a \leqq x \leqq b$, $a \leqq t \leqq b$. $|K(x, t) \leqq M$ in R, $K(x, t) \not\equiv 0$.

c) $f(x) \not\equiv 0$, is real and continuous in I: $a \leqq x \leqq b$.

d) λ, constant.

then the equation (5) has one and only one continuous solution $u(x)$ in I, and this solution is given by the absolutely and uniformly convergent series (6′).

The results of this article hold without change for the equation

$$u(x) = f(x) + \int_a^x K(x, t)u(t)dt$$

by putting throughout the discussion $\lambda = 1$.

10. Successive Approximations.—We would like to point out that the method of solution by successive approximations differs from that of successive substitutions.

Under the method of successive approximations we select any real function $u_o(x)$ continuous in I. Substitute in the right-hand member of

$$u(x) = f(x) + \lambda \int_a^b K(x,t)\, u(t)dt, \tag{1}$$

in place of $u(t)$, the function $u_o(t)$. We find

$$u_1(x) = f(x) + \lambda \int_a^b K(x, t)u_o(t)dt.$$

The function $u_1(x)$ so determined is real and continuous in I. Continue in like manner by replacing u_o by u_1, and so on. We obtain a series of functions

$$u_o(x),\ u_1(x),\ u_2(x),\ \ldots,\ u_n(x),\ \ldots$$

which satisfy the equations,

$$\left\{\begin{aligned} u_2(x) &= f(x) + \lambda \int_a^b K(x, t)u_1(t)dt \\ &\cdots\cdots\cdots\cdots \\ u_{n-1}(x) &= f(x) + \lambda \int_a^b K(x, t)u_{n-2}(t)dt \\ u_n(x) &= f(x) + \int_a^b K(x, t)u_{n-1}(t)dt \\ &\cdots\cdots\cdots\cdots \end{aligned}\right. \tag{7}$$

From these equations we find

$$(8)\quad u_n(x) = f(x) + \lambda \int_a^b K(x, t) f(t) dt$$
$$+ \lambda^2 \int_a^b K(x, t) \int_a^b K(t, t_1) f(t_1) dt_1 dt + \cdots$$
$$+ \lambda^{n-1} \int_a^b K(x, t) \int_a^b K(t, t_1) \cdots$$
$$\int_a^b K(t_{n-3}, t_{n-2}) f(t_{n-2}) dt_{n-2} \cdots dt_1 dt + R_n$$

where

$$R_n = \lambda^n \int_a^b K(x, t) \int_a^b K(t, t_1) \cdots \int_a^b K(t_{n-2}, t_{n-1})$$
$$u_o(t_{n-1}) dt_{n-1} \cdots dt_1 dt$$

$u_o(x)$ is real and continuous in I and so has a maximum value U in I. Then it is easy to see that

$$|R_n| \leq |\lambda^n| U M^n (b - a)^n.$$

If, then, $|\lambda| M(b - a) < 1$, we have

$$\lim_{n \to \infty} R_n = 0.$$

Thus, as n increases, the series of functions $u_n(x)$ approach a limit function which is given by the series in the right member of (8). We identify this series with the right member of (6′). Thus

$$\lim_{n \to \infty} u_n(x) \equiv u(x).$$

By this process at each step a new function $u_n(x)$ appears dependent upon the choice of $u_o(x)$. We notice, however, that the limit $u(x)$ is independent of the choice of $u_o(x)$.

We can now make an independent proof of the uniqueness of the solution. Suppose there was another solution $v(x)$. Choose $u_o(x) \equiv v(x)$. It is then clear that each $u_n(x)$ will be identical with $v(x)$ and hence the limit will be

$v(x)$. But we have just seen that the limit is independent of the choice of $u_0(x)$. Therefore,

$$v(x) \equiv u(x).$$

A similar discussion can be carried through without further difficulty for the Volterra equation.

11. Iterated Functions.—Place

$$(9) \qquad \begin{cases} K_1(x, t) = K(x, t) \\ K_i(x, t) = \int_a^b K(x, s)K_{i-1}(s, t)ds. \end{cases}$$

The functions $K_1, K_2, \ldots, K_n, \ldots$ so formed are called *iterated functions.*

By successive applications of (9) it is evident that

$$(10) \quad K_i(x, t) = \int_a^b \cdots \int_a^b K(x, s_1)K(s_1, s_2) \cdots K(s_{i-1}, t)ds_{i-1} \cdots ds_1.$$

From (10) $K_n(x, s)$ is an $(n - 1)$-fold integral and $K_p(s, t)$ is a $(p - 1)$-fold integral. Whence we see that $\int_a^b K_n(x, s)K_p(s, t)ds$ is an $(n + p - 1)$-fold integral, which, by some simple changes in the order of integration, is seen to be identical with $K_{n+p}(x, t)$. Hence

$$(11) \qquad K_{n+p}(x, t) = \int_a^b K_n(x, s)K_p(s, t)ds.$$

12. Reciprocal Functions.—Let

$$(12) \quad -k(x, t) = K_1(x, t) + K_2(x, t) + \cdots + K_n(x, t) + \cdots$$

It is easy to show that, when $K(x, t)$ is real and continuous in R, the infinite series for $k(x, t)$ is absolutely and uniformly convergent if $M(b - a) < 1$. Consequently, $k(x, t)$ is real and continuous in R. On account of the first of equations (9) and equation (11), we have

$$- k(x, t) - K(x, t) = K_2(x, t) + K_3(x, t) + \dots + K_n(x, t) + \dots$$

$$= \int_a^b K_1(x, s)K_1(s, t)ds + \dots + \int_a^b K_1(x, s)K_{n-1}(s, t)ds + \dots$$

$$= \int_a^b K_1(x, s)K_1(s, t)ds + \dots + \int_a^b K_{n-1}(x, s)\, K_1(s, t)ds + \dots$$

These equations may be written

$$- k(x, t) - K(x, t) = \int_a^b K_1(x, s)\left[K_1(s, t) + \dots + K_{n-1}(s, t) + \dots \right]ds$$

$$= \int_a^b \left[K_1(x, s) + \dots + K_{n-1}(x, s) + \dots \right] K_1(s, t)ds.$$

If we now make use of (12), we obtain the following characteristic formula:

$$\text{(13)} \qquad K(x, t) + k(x, t) = \int_a^b K(x, s)k(s, t)ds = \int k(x, s)K(s, t)ds.$$

Two functions $K(x, t)$ and $k(x, t)$ are said to be *reciprocal* if they are both real and continuous in R and if they satisfy the condition (13). A function $k(x, t)$ reciprocal to $K(x, t)$ will exist, provided the series in (12) converges uniformly. But we have seen that this series converges uniformly when $M(b - a) < 1$, where M is the maximum of $|K(x, t)|$ in R. Thus, we have the

Theorem III.—*If $K(x, t)$ is real and continuous in R, there exists a reciprocal function $k(x, t)$ given by* (12) *provided that*

$$M(b - a) < 1$$

where M is the maximum of $|K(x, t)|$ in R.

13. Volterra's Solution of Fredholm's Equation.—Volterra has shown how to find a solution of

$$u(x) = f(x) + \int_a^b K(x, t)u(t)dt \tag{4}$$

whenever the reciprocal function $k(x, t)$ of $K(x, t)$ is known. If (4) has a continuous solution $u(x)$, then

$$u(t) = f(t) + \int_a^b K(t, t_1)u(t_1)dt_1.$$

Multiplying by $k(x, t)$ and integrating, we find

$$\begin{aligned}\int_a^b k(x, t)u(t)dt &= \int_a^b k(x, t)f(t)dt \\ &\quad + \int_a^b \int_a^b k(x, t)K(t, t_1)u(t_1)dt_1dt \\ &= \int_a^b k(x, t)f(t)dt \\ &\quad + \int_a^b \left[K(x, t_1) + k(x, t_1)\right]u(t_1)dt_1,\end{aligned}$$

which reduces to

$$0 = \int_a^b k(x, t)f(t)dt + \int_a^b K(x, t_1)u(t_1)dt_1. \tag{14}$$

But from (4) we have

$$\int_a^b K(x, t_1)u(t_1)dt_1 = u(x) - f(x).$$

Therefore, (14) may be written

$$u(x) = f(x) - \int_a^b k(x, t)f(t)dt. \tag{15}$$

If (4) has a continuous solution, it is given by this formula and it is unique.

To see that the expression for $u(x)$ given by (15) is, indeed, a solution, we write (15) in the form

$$f(x) = u(x) + \int_a^b k(x, t)f(t)dt.$$

This is an integral equation for the determination of $f(x)$. The function reciprocal to $k(x, t)$ is $K(x, t)$. By what we have just proved, if this equation has a continuous solution, it is unique and is given by

$$f(x) = u(x) - \int_a^b K(x, t)u(t)dt.$$

But this is the equation (1) from which we started. Thus we see that (4) is satisfied by the value of $u(x)$ given by (15). Thus we have the following

Theorem IV.—*If*

a) $K(x, t)$ *is real and continuous in* R. $K(x, t) \not\equiv o$.

b) $f(x)$ *is real and continuous in* I. $f(x) \not\equiv o$.

c) A function $k(x, t)$ *reciprocal to* $K(x, t)$ *exists, then the equation* (4) *has one and only one continuous solution in* I *and this solution is given by* (15).

The same reasoning applied to (13), considered as an integral equation for the determination of $k(x, t)$, shows that, if a continuous reciprocal function exists, it is unique.

14. Discontinuous Solutions.—We have shown the existence, under proper assumptions, of a unique continuous solution for a linear integral equation of the second kind. This integral equation may have also, in addition, discontinuous solutions. To show this we exhibit the special equation[1]

$$u(x) = \int_o^x t^{x-t}u(t)dt,$$

which has one and only one continuous solution, namely $u(x) \equiv 0$. We can show, by direct substitution, that this

[1] Bôcher, "Integral Equations," p. 17, Cambridge Press, 1909.

equation has also an infinite number of discontinuous solutions given by

$$u(x) = Cx^{x-1},$$

where C is an arbitrary constant not zero.

EXERCISES

Solve the following linear integral equations:

1. *a*) $u(x) = x + \int_0^x (t - x)u(t)dt.$ *Ans.* $u(x) = \sin x.$

b) $u(x) = 1 + \int_0^x (t - x)u(t)dt.$ *Ans.* $u(x) = \cos x.$

2. $u(x) = \frac{5x}{6} + \frac{1}{2}\int_0^1 xt.u(t)dt.$ *Ans.* $u(x) = x.$

3. $u(x) = \frac{5x}{6} - \frac{1}{9} + \frac{1}{3}\int_0^1 (t + x)u(t)dt.$ *Ans.* $u(x) = x.$

4. $u(x) = 1 + \int_0^x u(t)dt.$ *Ans.* $u(x) = e^x.$

5. $u(x) = e^x - \frac{e}{2} + \frac{1}{2} + \frac{1}{2}\int_0^1 u(t)dt.$ *Ans.* $u(x) = e^x.$

6. $u(x) = \sin x - \frac{x}{4} + \frac{1}{4}\int_0^{\frac{\pi}{2}} tx\, u(t)dt.$ *Ans.* $u(x) = \sin x.$

7. $u(x) = x + \int_0^{1/2} u(t)dt.$ *Ans.* $u(x) = x + \text{constant}.$

8. $u(x) = 1 - 2x - 4x^2 + \int_0^x [3 + 6(x - t) - 4(x - t)^2]u(t)dt.$

Ans. $u(x) = e^x.$

9. $u(x) = \frac{3}{2}e^x - \frac{xe^x}{2} - \frac{1}{2} + \frac{1}{2}\int_0^1 t\, u(t)dt.$ *Ans.* $u(x) = e^x.$

10. $u(x) = 29 + 6x + \int_0^x (6x - 6t + 5)u(t)dt.$

Ans. $u(x) = e^{2x} - e^{3x}.$

11. $u(x) = \cos x - x - 2 + \int_0^x (t - x)u(t)dt.$

Ans. $u(x) = \sin x + x \sin x.$

12. Using the method of successive approximations find five successive approximations in the solution of Exercises 1, 2, 3, 4, 5 after choosing $u_o(x) \equiv 0$.

13. Show that

$$u(x) = A + Bx + \int_o^x [C + D(x - t)]u(t)dt$$

where A, B, C, D are arbitrary constants, has for solution

$$u(x) = K_1e^{m_1x} + K_2e^{m_2x},$$

where K_1, K_2, m_1, m_2 depend upon A, B, C, D.

14. Show that

$$u(x) = f(x) + \lambda\int_a^b u(t)\left[\sum_1^p \alpha_q(t)\beta_q(t)\right]dt$$

has the solution

$$u(x) = f(x) + \lambda\sum_1^p A_q\alpha_q(x),$$

where the A_q are constants determined by the equation

$$\sum_1^p rA_r\left[\lambda\int_a^b \alpha_r(t)\beta_q(t)dt\right] - A_q = \int_a^b \beta_q(t)f(t)dt \quad (q = 1, \ldots, p).$$

[Heywood-Frèchet].

CHAPTER III

SOLUTION OF FREDHOLM'S EQUATION EXPRESSED AS RATIO OF TWO INTEGRAL SERIES IN λ

15. Fredholm's Equation As Limit of a Finite System of Linear Equations.—The solution given in the previous chapter for the equation

$$u(x) = f(x) + \lambda \int_a^b K(x,t)\, u(t)dt \tag{1}$$

has the disadvantage of holding only for restricted values of λ. It is desirable to have, if possible, a solution which holds for all values of λ. Such a solution was given by Fredholm in the form

$$u(x) = \frac{\beta_0(x) + \beta_1(x)\lambda + \quad . \quad . \quad .}{\alpha_0 + \alpha_1\lambda + \quad . \quad . \quad .},$$

the numerator and denominator being permanently converging power series in λ.

a) The System of Linear Equations Replacing the Integral Equation.—Before stating explicitly and proving Fredholm's result, we give an outline of the reasoning which led him to his discovery.

Divide the interval (ab) into n equal parts and call the points of division $t_1, t_2, \ . \ . \ . \ , t_{n-1}$. Then

$$t_0 = a,\ t_1 = a + h,\ t_2 = a + 2h,\ \ . \ . \ . \ ,\ t_n = a + nh, \qquad h = \frac{b - a}{n}. \tag{2}$$

Replace the definite integral in (1) by the sum, corresponding to the points of division (2), of which it is the limit. We obtain the approximate equation

$$u(x) - \lambda h[K(x, t_1)u(t_1) + K(x, t_2)u(t_2) + \dots + K(x, t_n)u(t_n)] = f(x).$$

Since this equation holds for every value of x, it must be satisfied for $x = t_1, t_2, \dots, t_n$. We thus obtain the following system of n linear equations for the determination of the n unknowns $u(t_1), u(t_2), \dots, u(t_n)$:

$$(3)\quad \begin{aligned} u(t_1) - \lambda h[K(t_1, t_1)u(t_1) + K(t_1, t_2)u(t_2) + \dots + K(t_1, t_n)u(t_n)] &= f(t_1) \\ u(t_2) - \lambda h[K(t_2, t_1)u(t_1) + K(t_2, t_2)u(t_2) + \dots + K(t_2, t_n)u(t_n)] &= f(t_2) \\ \dots\dots\dots\dots\dots\dots\dots\dots \\ u(t_n) - \lambda h[K(t_n, t_1)u(t_1) + K(t_n, t_2)u(t_2) + \dots + K(t_n, t_n)u(t_n)] &= f(t_n). \end{aligned}$$

This system being solved with respect to $u(t_1), u(t_2), \dots,$ $u(t_n)$, we can plot $u(t_i)$ $(i = 1, \dots, n)$ as ordinates and by interpolation draw a curve $u(x)$, which we may expect

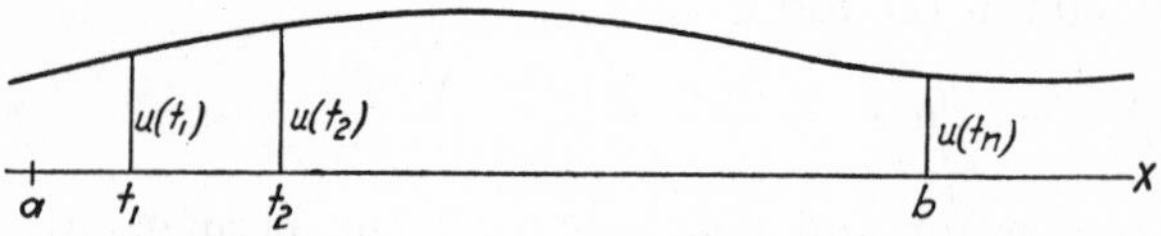

FIG. 2.

to be an approximation to the actual solution. We solve (3), making use of the abbreviations

$$f(t_i) = f_i \quad , \quad u(t_i) = u_i \quad , \quad K(t_i, t_j) = K_{ij}.$$

Denote by Δ the determinant of the coefficients of the u_i in (3). We have

$$\Delta = \begin{vmatrix} 1 - \lambda h K_{11} & -\lambda h K_{12} & \dots & -\lambda h K_{1n} \\ -\lambda h K_{21} & 1 - \lambda h K_{22} & \dots & -\lambda h K_{2n} \\ \dots & \dots & \dots & \dots \\ \dots & \dots & \dots & \dots \\ \dots & \dots & \dots & \dots \\ -\lambda h K_{n1} & -\lambda h K_{n2} & \dots & 1 - \lambda h K_{nn} \end{vmatrix}.$$

Denote by $\Delta_{\nu\mu}$ the first minor of the element in the νth row and μth column of Δ. Then, from Cramer's formula, we obtain, by solving (3) with respect to u_k,

$$(4) \quad u_k = \frac{\sum_1^n {}^i f_i \Delta_{ik}}{\Delta}, \text{ provided } \Delta \neq 0,\ K = 1, \ldots, n.$$

b) Limit of Δ.—Expanding the determinant Δ, we obtain

$$\Delta = 1 - \lambda \sum_{i=1}^{n} K_{ii} h + \frac{\lambda^2}{2!} \sum_{i,j=1}^{n} h^2 \begin{vmatrix} K_{ii} & K_{ij} \\ K_{ji} & K_{jj} \end{vmatrix} + \cdots$$

$$+ (-1)^n h^n \lambda^n \begin{vmatrix} K_{11} & \ldots & K_{1n} \\ \cdot & \cdots & \cdot \\ \cdot & \cdots & \cdot \\ K_{n1} & \ldots & K_{nn} \end{vmatrix}.$$

If we now let n increase indefinitely, we see that each term of this series has a definite limit. So that, at least formally,

$$\lim_{n \to \infty} \Delta = 1 - \lambda \int_a^b K(t, t) dt$$

$$+ \frac{\lambda^2}{2!} \int_a^b \int_a^b \begin{vmatrix} K(t_1, t_1) & K(t_1, t_2) \\ K(t_2, t_1) & K(t_2, t_2) \end{vmatrix} dt_1 dt_2$$

$$- \frac{\lambda^3}{3!} \int_a^b \int_a^b \int_a^b \begin{vmatrix} K(t_1, t_1) & \ldots & K(t_1, t_3) \\ \cdot & \cdots & \cdot \\ K(t_3, t_1) & \ldots & K(t_3, t_3) \end{vmatrix} dt_1 dt_2 dt_3 + \cdots$$

$$(5) \quad \equiv D(\lambda).$$

$D(\lambda)$ is called *Fredholm's determinant*, or the determinant of K.

c) Limit of Δ_{ik}.—The expression for $\Delta_{\mu\mu}$ is similar to that for Δ.

$$\Delta_{\mu\mu} = 1 - \lambda \sum_{i=1}^{n}{}' K_{ii} h + \frac{\lambda^2}{2!} \sum_{i,j=1}^{n}{}' h^2 \begin{vmatrix} K_{ii} & K_{ij} \\ K_{ji} & K_{jj} \end{vmatrix} + \cdots,$$

where ' means omit $i = \mu$. Hence

(6) $$\lim_{n\to\infty} \Delta_{\mu\mu} = D(\lambda).$$

Again, from rules for expansion of determinants,

$$\Delta_{\nu\mu} = \lambda h \left\{ K_{\mu\nu} - \lambda \sum_{i=1}^{n} h \begin{vmatrix} K_{\mu\nu} & K_{\mu i} \\ K_{i\nu} & K_{ii} \end{vmatrix} + \frac{\lambda^2}{2!} \sum_{i,j=1}^{n} h_2 \begin{vmatrix} K_{\mu\nu} & K_{\mu i} & K_{\mu j} \\ K_{i\nu} & K_{ii} & K_{ij} \\ K_{j\nu} & K_{ji} & K_{jj} \end{vmatrix} + \cdots \right\}.$$

Put $h\mathfrak{D}_{\mu\nu} = \Delta_{\nu\mu}$.

If, as n increases indefinitely, we let $(t_\mu t_\nu)$ vary in such a way that $\lim (t_\mu, t_\nu) = (x, y)$, we find, at least formally

(7) $$\lim_{n\to\infty} \mathfrak{D}_{\mu\nu} = \lambda K(x, y) - \lambda^2 \int_a^b \begin{vmatrix} K(x, y) & K(x, t) \\ K(t, y) & K(t, t) \end{vmatrix} dt$$
$$+ \frac{\lambda^3}{2!} \int_a^b \int_a^b \begin{vmatrix} K(x, y) & K(x, t_1) & K(x, t_2) \\ K(t_1, y) & K(t_1, t_1) & K(t_1, t_2) \\ K(t_2, y) & K(t_2, t_1) & K(t_2, t_2) \end{vmatrix} dt_1 dt_2 + \cdots$$

(8) $$\equiv D(x, y; \lambda).$$

This expression for $D(x, y; \lambda)$ is called *Fredholm's first minor.*

d) Limit of u_k.—We can now write (4) in the form

$$u_k = f_k \frac{\Delta_{kk}}{\Delta} + \sum_{i=1}^{n}{}' \frac{f_i \Delta_{ik}}{\Delta}$$
$$= f_k \frac{\Delta_{kk}}{\Delta} + \sum_{i=1}^{n}{}' \frac{f_i h D_{ki}}{\Delta},$$

which in the passage to the limit as $n \to \infty$, becomes

$$u(t_k) = f(t_k) + \frac{1}{D(\lambda)} \int_a^b f(t) D(t_k, t; \lambda) dt.$$

But t_k is any point of division. Then we can replace t_k by x and write

(9) $$u(x) = f(x) + \frac{1}{D(\lambda)} \int_a^b f(t) D(x, t; \lambda) dt.$$

This result has not been obtained by a rigorous mathematical procedure. However, we are inclined to believe that the above expression for $u(x)$ is a solution of (1). This belief is later shown to be correct.

e) Fredholm's Two Fundamental Relations.—We will now develop two relations which will be of use to us later in obtaining a solution of the integral equation (1). We recall the theorem, fundamental in the theory of determinants, that the sum of the products of the elements of any *column* by the corresponding minors of any other column is zero. This theorem applied to the determinant Δ gives

$$(1 - \lambda h K_{jj})\Delta_{jk} - \lambda h K_{kj}\Delta_{kk} - \sum_{i=1}^{n}{}'' \lambda h K_{ij}\Delta_{ik} = 0,$$

where $''$ means omit $i = j, k$. Making use of the relation $\Delta_{\nu\mu} = h\mathfrak{D}(t\mu, t\nu)$, we find

$$(1 - \lambda h K_{jj})h\mathfrak{D}_{kj} - \lambda h K_{kj}\Delta_{kk} - \sum_{i=1}^{n}{}'' \lambda h^2 K_{ij}\mathfrak{D}_{ki} = 0.$$

Divide through by h, since $h \neq 0$. The passage to the limit, as $n \to \infty$, gives from the last equation by (6) and (7)

$$D(t_k, t_j; \lambda) - \lambda K(t_k, t_j)D(\lambda) - \lambda \int_a^b K(t, t_j)D(t_k, t; \lambda)dt = 0.$$

This last equation holds for any two points t_j, t_k on the interval (ab). Let us put then $t_k = x$, $t_j = y$ and write

$$(10)\quad D(x, y; \lambda) - \lambda K(x, y)D(\lambda) = \lambda \int_a^b K(t, y)D(x, t; \lambda)dt.$$

This is called Fredholm's *first fundamental relation.*

Now apply the theorem: The sum of the products of the elements of any *row* by the corresponding minors of any other row is zero. This theorem applied to the determinant Δ gives

$$(1 - \lambda h K_{jj})\Delta_{kj} - \lambda h K_{jk}\Delta_{kk} - \sum_{i=1}^{n}{}'' \lambda h K_{ji}\Delta_{ki} = 0.$$

Proceeding as before, we find

$$D(t_j, t_k; \lambda) - \lambda K(t_j, t_k)D(\lambda) - \lambda \int_a^b K(t_j, t)D(t, t_k; \lambda)dt = 0.$$

This equation holds for any two points t_j, t_k on (ab). Let us put then $t_j = x$, $t_k = y$ and write

$$(11) \quad D(x,y;\lambda) - \lambda K(x,y)D(\lambda) - \lambda \int_a^b K(x,t)D(t,y;\lambda)dt = 0.$$

This is Fredholm's *second fundamental relation.*

16. Hadamard's Theorem.—We now proceed to establish rigorously the results of the preceding article. To this end we need a theorem due to Hadamard. To establish this theorem we make use of the following

Lemma.—If all of the elements a_{ik} of the determinant

$$A = \begin{vmatrix} a_{11} & a_{12} & . & . & . & a_{1n} \\ a_{21} & a_{22} & . & . & . & a_{2n} \\ . & . & . & . & . & . \\ a_{n1} & a_{n2} & . & . & . & a_{nn} \end{vmatrix}$$

are real and satisfy the conditions

$$(12) \quad a^2_{r1} + a^2_{r2} + \ . \ . \ . \ + a^2_{rn} = 1 \quad (r = 1, \ . \ . \ . \ , n),$$

then

$$|A| \leqq 1.$$

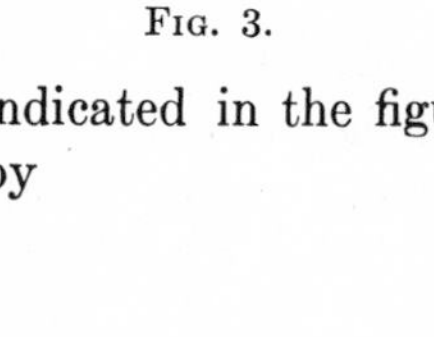

FIG. 3.

We give first two special cases of the lemma which have a geometric interpretation.

1) $n = 2$. The parallelogram $OP_1P_3P_2$ has the vertex O at the origin of a system of rectangular coordinates. The coordinates of P_1 and P_2 are as indicated in the figure. The area A of $OP_1P_3P_2$ is given by

$$A = \begin{vmatrix} x_1 & y_1 \\ x_2 & y_2 \end{vmatrix}.$$

If $OP_1 = OP_2 = 1$, that is, if

$$x_1^2 + y_1^2 = 1 \text{ and } x_2^2 + y_2^2 = 1,$$

then it is geometrically evident that the greatest area is obtained when the figure is a rectangle, and then the area is 1. Hence, we have generally $|A| \leqq 1$.

2) $n = 3$. The parallelopiped $OP_1P_2P_3$ has one vertex at the origin of a system of rectangular coordinates. The coordinates of P_1, P_2, P_3 are as indicated in the figure. The volume V of $OP_1P_2P_3$ is given by

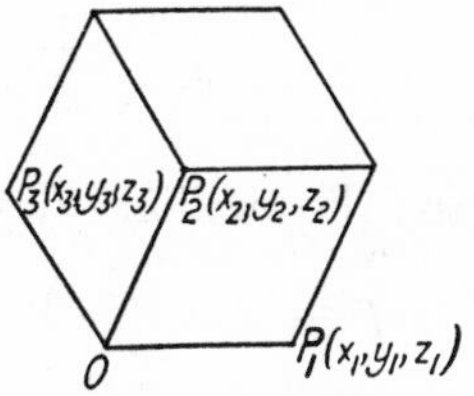

FIG. 4.

$$V = \begin{vmatrix} x_1 & y_1 & z_1 \\ x_2 & y_2 & z_2 \\ x_3 & y_3 & z_3 \end{vmatrix}.$$

If $OP_1 = OP_2 = OP_3 = 1$, that is, if
$x_1^2 + y_1^2 + z_1^2 = 1, x_2^2 + y_2^2 + z_2^2 = 1, x_3^2 + y_3^2 + z_3^2 = 1,$
then it is evident geometrically that the volume is greatest when the figure is a rectangular parallelopiped, in which case the volume is 1. Hence, we have generally

$$|V| \leqq 1.$$

Proof of Lemma.—$A(a_{11}, \ . \ . \ . \ , a_{nn})$ is a continuous function of its arguments a_{rs} in the region $\mathfrak{A}$ defined by the equations (12). These conditions insure that $|a_{rs}| \leqq 1$, and that the region $\mathfrak{A}$ is bounded and closed. Hence A reaches a maximum and a minimum on the region $\mathfrak{A}$. The maximum and minimum which we are seeking are the so-called absolute maximum and minimum. But if a system of values furnishes the absolute maximum (minimum) for A, it furnishes also a relative maximum (minimum). Hence the ordinary methods of the differential calculus can be used for their determination.

Now if a function

$$f(x_1, x_2, \ . \ . \ . \ , x_n)$$

of n variables, connected by h distinct relations

$$\phi_1(x_1, \ . \ . \ . \ , x_n) = 0, \quad \phi_2(x_1, \ . \ . \ . \ , x_n) = 0, \ . \ . \ . \ , \quad \phi_h(x_1, \ . \ . \ . \ , x_n) = 0$$

has a maximum (minimum), then the n first partial derivatives of the auxiliary function

$$F = f + \lambda_1\phi_1 + \lambda_2\phi_2 + \ . \ . \ . \ + \lambda_h\phi_h,$$

where $\lambda_1, \lambda_2, \ . \ . \ . \ , \lambda_h$ are constants, must vanish.[1]

For our problem, $f = A,\ x_r = a_{rs}$

$$\phi_r = \sum_{s=1}^{n} a_{rs}^2 - 1 \quad (r = 1, \ . \ . \ . \ , n)$$

and the auxiliary function F becomes

$$F = A + \sum_{s=1}^{n} \frac{\lambda_s}{2}(a_{s1}^2 + a_{s2}^2 + \ . \ . \ . \ + a_{sn}^2 - 1).$$

For a maximum (minimum) by the theorem just stated, we must have

$$\frac{\partial F}{\partial a_{jk}} = \frac{\partial A}{\partial a_{jk}} + \lambda_j a_{jk} = 0, \text{ or}$$

$$(13) \qquad A_{jk} + \lambda_j a_{jk} = 0, \quad (j,\, k = 1, 2, \ . \ . \ . \ , n),^2$$

where A_{jk} denotes the minor of a_{jk} in A. Multiply both sides of this equation by a_{jk} and sum with respect to k for $k = 1, 2, \ . \ . \ . \ , n$. We obtain

$$A + \lambda_j = 0, \quad \text{since} \sum_{k=1}^{n} a_{jk}^2 = 1$$

or

$$\lambda_j = -A \quad (j = 1, \ . \ . \ . \ , n).$$

Substitute this value of λ_j in equation (13). Then

$$A_{jk} = Aa_{jk} \qquad (jk = 1, \ . \ . \ . \ , n).$$

Whence the determinant

$$\begin{vmatrix} A_{11} & A_{12} & . \ . \ . & A_{1n} \\ A_{21} & A_{22} & . \ . \ . & A_{2n} \\ . & . & . \ . \ . & . \\ A_{n1} & A_{n2} & . \ . \ . & A_{nn} \end{vmatrix}$$

[1] GOURSAT-HEDRICK, "Mathematical Analysis," vol. 1, §61, Ginn & Co.

[2] For differentiation of determinants, see BALTZER, "Die Determinanten," §3, 14.

adjoint to A, is equal to

$$\begin{vmatrix} Aa_{11} & Aa_{12} & \cdots & Aa_{1n} \\ Aa_{21} & Aa_{22} & \cdots & Aa_{2n} \\ \cdots & \cdots & \cdots & \cdots \\ Aa_{n1} & Aa_{n2} & \cdots & Aa_{nn} \end{vmatrix}.$$

The first of these determinants is equal[1] to A^{n-1}. The second reduces to A^{n+1}. Hence

$$A^{n+1} = A^{n-1}$$

But the maximum and minimum of A must satisfy this equation. Therefore, the maximum of A is $+1$, the minimum is -1, and

$$|A| \leqq 1.$$

Hadamard's Theorem.—We are now in a position to prove a more general theorem due to Hadamard: *If the elements* b_{ik} *of the determinant*

$$B = \begin{vmatrix} b_{11} & b_{12} & \cdots & b_{1n} \\ b_{21} & b_{22} & \cdots & b_{2n} \\ \cdots & \cdots & \cdots & \cdots \\ b_{n1} & b_{n2} & \cdots & b_{nn} \end{vmatrix}$$

are real and satisfy the inequality

$$|b_{ik}| \leqq M,$$

then

$$|B| \leqq M^n\sqrt{n^n}.$$

Proof.—Let

$$b_{i1}^2 + b_{i2}^2 + \cdots + b_{in}^2 = s_i, \quad (i = 1, \ldots, n)$$

Case I.—Some one or more of the s_i *vanish, say* $s_k = 0$. Then $b_{ki} = 0$ $(i = 1, \ldots, n)$. Therefore $B = 0$, and the theorem is proved in this case.

Case II.—None of the s_i vanish. Then each s_i is positive. That is

$$s_1 > 0, s_2 > 0, \ldots, s_n > 0.$$

[1] BALTZER, *Loc. cit.*, §6, 1.

We now consider the determinant

$$\frac{B}{\sqrt{s_1 s_2 \ . \ . \ . \ s_n}} = \begin{vmatrix} \frac{b_{11}}{\sqrt{s_1}} & \cdots & \frac{b_{1n}}{\sqrt{s_1}} \\ \cdot & \cdots & \cdot \\ \cdot & \cdots & \cdot \\ \frac{b_{n1}}{\sqrt{s_n}} & \cdots & \frac{b_{nn}}{\sqrt{s_n}} \end{vmatrix}$$

for which $\frac{b_{i1}^2}{s_i} + \ . \ . \ . \ + \frac{b_{in}^2}{s_i} = 1. \quad (i = 1, \ . \ . \ . \ , n).$

This determinant satisfies all of the conditions of our lemma. Hence

$$| B | \leqq \sqrt{s_1 s_2 \ . \ . \ . \ s_n}.$$

But, since $| b_{ik} | \leqq M,$

we have from $s_i = b_{i1}^2 + \ . \ . \ . \ + b_{in}^2$

$$s_i \leqq nM^2. \quad (i = 1, \ . \ . \ . \ , n)$$

Therefore, $| B | \leqq M^n \sqrt{n^n}.$

17. Convergence Proof.—With the help of Hadamard's theorem we can now prove the convergence of the series for $D(\lambda)$ and $D(x, y; \lambda)$.

a) Convergence of $D(\lambda)$.—$D(\lambda)$ is given by the series

$$(14) \qquad D(\lambda) = 1 + \sum_{n=1}^{\infty} (-1)^n \frac{\lambda^n}{n!} A_n, \text{ where}$$

$$(15) \quad A_n = \int_a^b \cdots \int_a^b \begin{vmatrix} K(t_1 t_1) . . K(t_1 t_n) \\ \cdots\cdots\cdots \\ \cdots\cdots\cdots \\ K(t_n t_1) . . K(t_n t_n) \end{vmatrix} dt_1 . . . dt_n \ (n > 0).$$

We have assumed in §8 that $| K(x, t) | \leqq M$ in R. Thus the determinant in the expression for A_n satisfies all of the conditions of Hadamard's theorem, and hence

$$| A_n | \leqq \int_a^b \ . \ . \ . \ \int_a^b M^n \sqrt{n^n} \, dt_1 \ . \ . \ . \ dt_n$$

$$= \sqrt{n^n} M^n (b - a)^n.$$

Then

$$\left|(-1)^n \frac{\lambda^n}{n!} A_n\right| \leq M^n (b-a)^n |\lambda^n| \frac{\sqrt{n^n}}{n!} \equiv C_n.$$

This identity defines C_n.

We now proceed to show that the series of which C_n is the general term converges. Applying the ratio test, we obtain

$$\frac{C_{n+1}}{C_n} = M(b-a)|\lambda| \sqrt{\left(1+\frac{1}{n}\right)^n} \frac{1}{\sqrt{n+1}}.$$

This ratio has for its limit zero as n becomes infinite. Hence the series of which C_n is the general term converges. Consequently, the series for $D(\lambda)$ is absolutely and permanently convergent. We state this result in the following:

Theorem I.—*The series $D(\lambda)$ is an absolutely and permanently converging power series in λ.*

b) Convergence of $D(x, y; \lambda)$.—We have

$$(16)\quad D(x, y; \lambda) = \lambda K(x, y) + \sum_{n=1}^{\infty} (-1)^n \frac{\lambda^{n+1}}{n!} B_n(x, y), \text{ where}$$

$$(17)\quad B_n(x, y) = \int_a^b \cdots \int_a^b \begin{vmatrix} K(x, y) & K(x, t_1) & \cdots & K(x, t_n) \\ K(t_1, y) & K(t_1, t_1) & \cdots & K(t_1, t_n) \\ \cdots & \cdots & \cdots & \cdots \\ K(t_n, y) & K(t_n, t_1) & \cdots & K(t_n, t_n) \end{vmatrix} dt_1 \ldots dt_n.$$

It is sometimes convenient to write

$$D(x, y; \lambda) = \sum_{n=0}^{\infty} (-1)^n \frac{\lambda^{n+1}}{n!} B_n(x, y),$$

where we consider $B_0(x, y) = K(x, y)$ and $0! = 1$. The determinant in the expression for B_n satisfies all of the conditions of Hadamard's theorem and hence

$$|B_n| \leq \int_a^b \cdots \int_a^b \sqrt{(n+1)^{n+1}} M^{n+1} dt_1 \ldots dt_n = \sqrt{(n+1)^{n+1}} M^{n+1} (b-a)^n$$

Then

$$\left|(-1)^n\frac{\lambda^{n+1}}{n!}B_n\right| \leqq M^{n+1}(b-a)^n|\lambda|^{n+1}\frac{\sqrt{(n+1)^{n+1}}}{n!} \equiv E_n.$$

This identity defines E_n. We now proceed to show that the series of which E_n is the general term converges. Applying the ratio test,

$$\frac{E_n}{E_{n-1}} = M(b-a)|\lambda|\sqrt{\left(1+\frac{1}{n}\right)^n}\cdot\frac{1}{\sqrt{n+1}}\cdot\frac{n+1}{n}.$$

The ratio has for its limit zero as n becomes infinite. Hence the series of which E_n is the general term converges. Consequently, the series for $D(x, y; \lambda)$ is absolutely and permanently convergent in λ and, moreover, uniformly convergent in x and y on R. Hence the

Theorem II.—*The series $D(x, y; \lambda)$ converges absolutely and permanently in λ, and, moreover, uniformly*[1] *on R: $a \leqq x \leqq b$, $a \leqq y \leqq b$.*

18. Fredholm's Two Fundamental Relations.—We will now proceed to prove the two relations:

$$(18)\quad D(x, y; \lambda) - \lambda K(x, y)D(\lambda) = \lambda\int_a^b D(x, t; \lambda)K(t, y)dt$$

$$(19)\quad = \lambda\int_a^b K(x, t)D(t, y; \lambda)\,dt$$

wihch were previously heuristically obtained.

a) Relation between the Coefficients $A_n(x, y)$ and $B_n(x, y)$.—Substitute in (18), in place of $D(x, y; \lambda)$ and $D(\lambda)$, their series expressions. Then both sides of the equality in (18) become power series in λ. Hence, if (18) is true, the coefficients of corresponding powers of λ on the two sides must be equal. Conversely, if we can show that the coefficients of corresponding powers of λ on the two sides are equal, then (18) will be established. Making the substitutions, we obtain

[1] GOURSAT-HEDRICK, "Mathematical Analysis," §173, note.

$$\lambda K(x, y) + \sum_{n=1}^{\infty} (-1)^n \frac{\lambda^{n+1}}{n!} B_n(x, y)$$

$$- \lambda K(x, y)\left[1 + \sum_{n=1}^{\infty} (-1)^n A_n \frac{\lambda^n}{n!}\right]$$

$$= \lambda \int_a^b K(t, y)\left[\lambda K(x, t) + \sum_{n=1}^{\infty} (-1)^n \frac{\lambda^{n+1}}{n!} B_n(x, t)\right] dt.$$

On the right-hand side, the series in the integrand is uniformly convergent and remains so when multiplied by $K(t, y)$. Therefore, we can integrate by terms and write

$$\lambda^2 \int_a^b K(x, t)K(t, y)dt + \sum_{n=1}^{\infty} (-1)^n \frac{\lambda^{n+2}}{n!} \int_a^b B_n(x, t)K(t, y)dt.$$

If in this second integral we put $n' = n + 1$ and then drop the prime, we obtain

$$\sum_{n=1}^{\infty} (-1)^n \frac{\lambda^{n+1}}{n!} B_n(x, y) - K(x, y) \sum_{n=1}^{\infty} (-1)^n \frac{\lambda^{n+1}}{n!} A_n$$

$$= \lambda^2 \int_a^b K(x, t)K(t, y)dt$$

$$+ \sum_{n=1}^{\infty} (-1)^{n-1} \frac{\lambda^{n+1}}{(n-1)!} \int_a^b B_{n-1}(x, t)K(t, y)dt.$$

Compare now the coefficients of corresponding powers of λ on the two sides. We obtain

$$(20) \quad B_n(x,y) = A_n \cdot K(x, y) - n \int_a^b B_{n-1}(x, t)K(t, y)dt.$$

If we establish the truth of this equality, then (18) will be shown to be true.

The relation (19) treated in exactly the same way leads to the relation

$$(21) \quad B_n(x, y) = A_n \cdot K(x, y) - n \int_a^b K(x, t)B_{n-1}(t, y)dt.$$

If we can establish the truth of this equality, then (19) will be shown to be true.

b) *Proof of* (20) *and* (21).—We prove (20) and also (21) by showing that the explicit expression for $B_n(x, y)$ can by proper expansion and change in notation be written in the form indicated.

The explicit expression for $B_n(x, y)$ is

$$B_n(x, y) = \int_a^b \cdots \int_a^b \begin{vmatrix} K(x, y) & K(x, t_1) & \cdots & K(x, t_n) \\ K(t_1, y) & K(t_1, t_1) & \cdots & K(t_1, t_n) \\ \cdots & \cdots & \cdots & \cdots \\ K(t_n, y) & K(t_n, t_1) & \cdots & K(t_n, t_n) \end{vmatrix} dt_1 \ldots dt_n.$$

Develop the determinant which appears in the integrand in terms of the elements of the first *column*. We obtain

$$B_n(x, y) = \int_a^b \cdots \int_a^b K(x, y) \times \begin{vmatrix} K(t_1, t_1) & \cdots & K(t_1, t_n) \\ \cdots & \cdots & \cdots \\ \cdots & \cdots & \cdots \\ K(t_n, t_1) & \cdots & K(t_n, t_n) \end{vmatrix} dt_1 \ldots dt_n$$

$$+\sum_{i=1}^{n} (-1)^i \int_a^b \cdots \int_a^b K(t_i, y) \times \begin{vmatrix} K(x, t_1) & \cdots & K(x, t_n) \\ K(t_1, t_1) & \cdots & K(t_1, t_n) \\ \cdots & \cdots & \cdots \\ \cdots & \cdots & \cdots \\ K(t_{i-1}, y) & \cdots & K(t_{i-1}, t_n) \\ K(t_{i+1}, y) & \cdots & K(t_{i+1}, t_n) \\ \cdots & \cdots & \cdots \\ \cdots & \cdots & \cdots \\ K(t_n, y) & \cdots & K(t_n, t_n) \end{vmatrix} dt_1 \ldots dt_n.$$

The first term on the right reduces to $K(x, y) \cdot A_n$, according to (15). In the terms which occur in the summation in place of

$$t_i,\ t_{i+1},\ t_{i+2},\ \ldots,\ t_n$$

write

$$t,\ t_i,\ t_{i+1},\ \ldots,\ t_{n-1},$$

which means simply a change in the notation of the variables of integration in a multiple definite integral. We obtain

$$\sum_{i=1}^{n}(-1)^i \int_a^b \cdots \int_a^b K(t, y) \times$$

$$\begin{vmatrix} K(x, t_1) & \ldots & K(x, t_{i-1})K(x, t)K(x, t_i) & \ldots & K(x, t_{n-1}) \\ \cdots & \cdots & \cdots & \cdots & \cdots \\ \cdots & \cdots & \cdots & \cdots & \cdots \\ \cdots & \cdots & \cdots & \cdots & \cdots \\ K(x, t_{n-1}) & \ldots & \ldots & \ldots & K(t_{n-1}\ t_{n-1}) \end{vmatrix} dt\, dt_1 \ldots dt_{n-1},$$

which may be written by bringing t into the first column

$$\sum_{i=1}^{n}(-1)^{2i-1} \int_a^b \cdots \int_a^b K(t, y) \times$$

$$\begin{vmatrix} K(x, t)\ K(x, t_1) & \ldots & K(x, t_{n-1}) \\ K(t_1, t)\ K(t_1, t_1) & \ldots & K(t_1, t_{n-1}) \\ \cdots & \cdots & \cdots \\ \cdots & \cdots & \cdots \\ K(t_{n-1}, t)\ K(t_{n-1}, t_1) & \ldots & K(t_{n-1}, t_{n-1}) \end{vmatrix} dt\, dt_1 \ldots dt_{n-1}.$$

This last expression shows that all of the n terms of the summation are equal. Furthermore, the integrations may be performed in any order. We then integrate first with respect to $t_1, \ldots, t_{n-1}$. For these integrations, $K(t, y)$ may be regarded as a constant factor and may be taken before the $(n - 1)$ – fold integral, so that we obtain

$$-n\int_a^b K(t, y)\left\{\int_a^b \cdots \int_a^b \begin{vmatrix} K(x, t) & K(x, t_1) & \cdots & K(x, t_{n-1}) \\ K(t_1, t) & K(t_1, t_1) & \cdots & K(t_1, t_{n-1}) \\ \cdots & \cdots & \cdots & \cdots \\ \cdots & \cdots & \cdots & \cdots \\ \cdots & \cdots & \cdots & \cdots \\ K(t_{n-1}, t) & K(t_{n-1}, t_1) & \cdots & K(t_{n-1}, t_{n-1}) \end{vmatrix} dt_1 \ldots dt_{n-1}\right\} dt,$$

which, according to (17), may be written

$$-n\int_a^b B_{n-1}(x, t)K(t, y)dt.$$

Hence the development of $B_n(x, y)$ gives

$$B_n(x, y) = A_n \cdot K(x, y) - n\int_a^b B_{n-1}(x, t)\, K(t, y)dt,$$

which proves relation (20). Therefore (18), which is Fredholm's first relation, is true.

Take again the explicit expression for $B_n(x, y)$, but this time develop the determinant in the integrand in terms of the elements of the first *row*. Then, proceeding as before, we obtain for the development of $B_n(x, y)$

$$B_n(x, y) = A_n \cdot K(x, y) - n\int_a^b K(x, t)B_{n-1}(t, y)dt,$$

which proves the relation (21). Therefore, (19), which is Fredholm's second relation, is true. Thus we have established the following.

Theorem III.—*Between Fredholm's determinant $D(\lambda)$ and Fredholm's first minor $D(x, y; \lambda)$ the following double relation holds:*

$$(18) \quad D(x, y; \lambda) - \lambda K(x, y)D(\lambda) = \lambda\int_a^b D(x, t; \lambda)\, K(t, y)dt$$

$$(19) \quad = \lambda\int_a^b K(x, t)D(t, y; \lambda)dt$$

for all values of λ and for all values of x and y on R.

19. Fredholm's Solution of the Integral Equation When $D(\lambda) \neq 0$.—Fredholm's two fundamental relations, the equations (18) and (19), enable us now to obtain a solution of the integral equation

$$(1) \qquad u(x) = f(x) + \lambda \int_a^b K(x, t)u(t)dt.$$

We obtain a hint as to the method of procedure from the method of solving the finite system

$$(3) \qquad u_i - \lambda h \sum_{j=1}^{n} K_{ij}u_j = f_i \quad (i = 1, \ . \ . \ . \ , n).$$

To find u_k from this system we first multiply by Δ_{ik} and sum with respect to i. We obtain

$$\sum_{i=1}^{n} u_i\Delta_{ik} - \lambda h \sum_{i=1}^{n} \sum_{j=1}^{n} K_{ij}u_j\Delta_{ik} = \sum_{i=1}^{n} f_i \, \Delta_{ik}$$

whence

$$(4) \qquad \Delta u_k = \sum_{i=1}^{n} f_i \, \Delta_{ik}.$$

Now

$$\Delta_{ik} = h\mathfrak{D}_{ki} \text{ by definition (see §15, } c)$$

and

$$\lim_{h \to 0} \sum_{i=1}^{n} h\mathfrak{D}_{ki} = \int_a^b D(x, t; \lambda)dt \text{ by (7).}$$

Let us now follow the analogy. Write (1) in the form

$$u(t) = f(t) + \lambda \int_a^b K(t, \xi)u(\xi)d\xi.$$

Multiply both sides of this equation, which we suppose is satisfied by a continuous function u, by $D(x, t; \lambda)$ and then integrate with respect to t from a to b. We obtain

$$(22) \quad \int_a^b D(x, t; \lambda)u(t)dt = \int_a^b D(x, t; \lambda)f(t)dt \\ + \lambda \int_a^b \int_a^b D(x, t; \lambda)K(t, \xi)u(\xi)d\xi dt.$$

The integrand of the double integral being continuous in x and ξ, we can interchange the order of integration[1] in the double integral and write it

$$\int_a^b u(\xi)\left[\lambda\int_a^b D(x, t; \lambda)K(t, \xi)dt\right]d\xi,$$

which, according to (18), becomes

$$\int_a^b\left[D(x, \xi; \lambda) - \lambda D(\lambda)K(x, \xi)\right]u(\xi)d\xi.$$

Hence (22) may be written

$$\int_a^b D(x, t; \lambda)u(t)dt = \int_a^b D(x, t; \lambda)f(t)dt$$
$$+ \int_a^b D(x, \xi; \lambda)\ u(\xi)d\xi - \lambda D(\lambda)\int_a^b K(x, \xi)u(\xi)d\xi,$$

which, on account of (1), reduces to

$$0 = \int_a^b D(x, t; \lambda)f(t)dt - D(\lambda)\left[u(x) - f(x)\right].$$

We solve now for $u(x)$, under the assumption $D(\lambda) \neq 0$. We obtain

$$(23)\qquad u(x) = f(x) + \int_a^b \frac{D(x, t; \lambda)f(t)}{D(\lambda)}dt.$$

Hence, if u is a continuous function of x which satisfies (1), and if $D(\lambda) \neq 0$, then $u(x)$ is given by (23).

It remains for us to show that also, conversely, the expression for $u(x)$ given by (23) is a solution of (1). We do this by direct substitution. Substitute the value of $u(x)$ as given by (23) in (1). We obtain

$$f(x) + \int_a^b \frac{D(x, t; \lambda)f(t)}{D(\lambda)}dt = f(x) + \lambda\int_a^b K(x, t) \times$$
$$\left\{f(t) + \int_a^b \frac{D(t, \xi; \lambda)f(\xi)}{D(\lambda)}d\xi\right\}dt.$$

[1] Goursat-Hedrick, "Mathematical Analysis," vol. 1, §123.

Break the last term up into two parts and in the double integral change the order of integration. We obtain

$$\int_a^b \frac{D(x, t; \lambda)f(t)}{D(\lambda)}dt = \lambda \int_a^b K(x,t)f(t)dt$$
$$+ \frac{1}{D(\lambda)}\int_a^b f(\xi)\left[\lambda \int_a^b K(x, t)D(t, \xi; \lambda)dt\right]d\xi,$$

which, according to (19), may be written

$$\int_a^b \frac{D(x, t; \lambda)f(t)}{D(\lambda)}dt = \lambda \int_a^b K(x, t)f(t)dt$$
$$+ \frac{1}{D(\lambda)}\int_a^b f(\xi)\Big[D(x, \xi; \lambda) - \lambda K(x, \xi)D(\lambda)\Big]d\xi.$$

But this last equation is seen to be an identity. Consequently, the expression for $u(x)$ given by (23) satisfies equation (1). Thus we have proved the following theorem, which is called *Fredholm's first fundamental theorem:*

Theorem IV.—*If*

(*a*) $D(\lambda) \neq 0.$

(*b*) *$K(x, t)$ is continuous in R.*

(*c*) *$f(x)$ is continuous in I.*

then the equation

$$(1) \qquad u(x) = f(x) + \lambda \int_a^b K(x, t)u(t)dt$$

has one and only one continuous solution given by

$$(23) \qquad u(x) = f(x) + \int_a^b \frac{D(x, t; \lambda)f(t)}{D(\lambda)}dt$$

where $D(x, t; \lambda)$ and $D(\lambda)$ are absolutely and permanently convergent integral series in λ, and $D(x, t; \lambda)$ converges uniformly with respect to x and t on R: $a \leq x \leq b$, $a \leq y \leq b$.

We have at once, for the special case $f \equiv 0$, the following:

Corollary.—If $D(\lambda) \neq 0$, then the homogeneous equation

$$(24) \qquad u(x) = \lambda \int_a^b K(x, t)u(t)dt$$

has one and only one continuous solution given by $u(x) \equiv 0$.

Let us point out the analogy with the finite system of linear equations

$$u_i - \lambda h \sum_{j=1}^{n} K_{ij}\, u_j = f_i \quad (i = 1, \ . \ . \ . \ , n)$$

with determinant Δ.

If $\Delta \neq 0$, then this system has one and only one solution.

If $f_j \equiv 0$, then the only solution is the trivial one

$$u_1 = \ . \ . \ . \ = u_n \equiv 0.$$

But the limit of Δ was $D(\lambda)$. Hence the results of Theorem IV and its corollary are exactly what was to be expected from the analogy with the finite system.

20. Solution of the Homogeneous Equation When $D(\lambda) = 0$, $D'(\lambda) \neq 0$.—The discussion up to this point has been made under the assumption $D(\lambda) \neq 0$. Let us now see what happens when $D(\lambda) = 0$, first with respect to the homogeneous equation (24).

Let λ_o be a value of λ for which

$$D(\lambda_o) = 0. \tag{25}$$

We now consider the solution of the homogeneous integral equation (24) for this particular value of λ:

$$u(x) = \lambda_o \int_a^b K(x, t)u(t)dt. \tag{26}$$

We obtain a solution of (26) by means of Fredholm's second fundamental relation (19), which is true for all values of λ and hence for $\lambda = \lambda_o$. With this value of λ, on account of (25), equation (19) becomes

$$D(x, y; \lambda_o) = \lambda_o \int_a^b K(x, t)D(t, y; \lambda_o)dt.$$

The equality holds for every value of y on the interval (ab) and, therefore, for $y = y_o$. Then

$$D(x, y_o; \lambda_o) = \lambda_o \int_a^b K(x, t)D(t, y_o; \lambda_o)dt.$$

But this is just the equation (26) with $u(x)$ replaced by $D(x, y_o; \lambda_o)$. Thus we see that $u(x) = D(x, y_o; \lambda_o)$ is a solution of (26). Moreover, this solution is continuous,[1] for $D(x, y; \lambda)$ is uniformly convergent in x and y and its terms are continuous. But $D(x, y_o; \lambda_o)$ may be identically zero in x, either on account of an unfortunate choice of y_o, in which case we could choose some other value for y_o, or because $D(x, y; \lambda_o) \equiv 0$ in x and y, in which case the above solution reduces to the trivial one $u \equiv 0$, no matter how we choose y_o. We have thus proved the following:

Theorem V.—*If* $D(\lambda_o) = 0$ *and* $D(x, y; \lambda_o) \not\equiv 0$, *then for a proper choice of* y_o, $u(x) = D(x, y_o; \lambda_o)$ *is a continuous solution of*

$$u(x) = \lambda_o \int_a^b K(x, t)u(t)dt$$

and $u(x) \not\equiv 0$.

In the theorem just stated, the condition $D(x, y; \lambda_o) \not\equiv 0$ may be replaced by the condition $D'(\lambda) \neq 0$. To show this we prove the following

$$\int_a^b D(x, x; \lambda)dx = -\lambda D'(\lambda). \tag{27}$$

We prove (27) by making use of the series expressions for $D'(\lambda)$ and $D(x, y; \lambda)$. We have from (14)

$$D(\lambda) = 1 - \lambda A_1 + \frac{\lambda^2}{2!}A_2 - \frac{\lambda^3}{3!}A_3 + \cdots$$

where the A_n are given by (15). Then

$$D'(\lambda) = -A_1 + \lambda A_2 - \frac{\lambda^2}{2!}A_3 + \cdots$$

$$= -\sum_{n=0}^{\infty}(-1)^n\frac{\lambda^n}{n!}A_{n+1}.$$

[1] GOURSAT-HEDRICK, "Mathematical Analysis," vol. 1, §173.

In the expression for A_{n+1}:

$$A_{n+1} = \int_a^b \cdots \int_a^b \begin{vmatrix} k(t_1, t_1)K(t_1, t_2) & \cdots & K(t_1, t_{n+1}) \\ K(t_2, t_1)K(t_2, t_2) & \cdots & K(t_2, t_{n+1}) \\ \cdots & \cdots & \cdots \\ K(t_{n+1}, t_1)K(t_{n+1}, t_2) & \cdots & K(t_{n+1}, t_{n+1}) \end{vmatrix} dt_1 \cdots dt_{n+1},$$

in place of $t_1, t_2, t_3, \ldots, t_n, \quad t_{n+1}$
put $x, t_1, t_2, \ldots, t_{n-1}, t_n.$

Then

$$A_{n+1} = \int_a^b \cdots \int_a^b \begin{vmatrix} K(x, x) \; K(x, t_1) & \cdots & K(x, t_n) \\ K(t_1, x) \; K(t_1, t_1) & \cdots & K(t_1, t_n) \\ \cdots & \cdots & \cdots \\ K(t_n, x) \; K(t_n, t_1) & \cdots & K(t_n, t_n) \end{vmatrix} dx dt_1 \cdots dt_n.$$

In this multiple definite integral we change the order of integration[1] and integrate first with respect to $dt_1 \ldots dt_n$, whence

$$A_{n+1} = \int_a^b \left\{ \int_a^b \cdots \int_a^b \begin{vmatrix} K(x, x) \; K(x, t_1) & \cdots & K(x, t_n) \\ K(t_1, x) \; K(t_1, t_1) & \cdots & K(t_1, t_n) \\ \cdots & \cdots & \cdots \\ K(t_n, x) \; K(t_n, t_1) & \cdots & K(t_n, t_n) \end{vmatrix} dt_1 \cdots dt_n \right\} dx,$$

which, on account of (17), becomes

$$A_{n+1} = \int^b B_n(x, x) dx.$$

[1] GOURSAT-HEDRICK "Mathematical Analysis," vol 1, §123; PIERPONT, JAMES, "The Theory of Functions of Real Variables," vol. 1, §570.

Therefore,

$$D'(\lambda) = -\sum_{n=0}^{\infty}(-1)^n \frac{\lambda^n}{n!}\int_a^b B_n(x, x)dx.$$

But

$$\sum_{n=0}^{\infty}(-1)^n \frac{\lambda^n}{n!} B_n(x, x)dx$$

is a series uniformly convergent in x. We can then interchange the order of writing the summation and the integration[1] in the expression for $D'(\lambda)$ and write

$$D'(\lambda) = -\int_a^b \sum_{n=0}^{\infty}(-1)^n \frac{\lambda^n}{n!} B_n(x, x)dx.$$

Multiply both sides by $-\lambda$, and then by (16) we see that, indeed,

$$\int_a^b D(x, x; \lambda)dx = -\lambda D'(\lambda). \tag{27}$$

Let us suppose now that $D(\lambda_o) = 0$ and $D'(\lambda_o) \neq 0$, then certainly $\lambda_o \neq 0$, since $D(0) = 1$. Hence, if we write (27) with $\lambda = \lambda_o$, the right-hand side of this equality is different from zero and so also the left-hand side, and, therefore, $D(x, x; \lambda_o) \not\equiv 0$ in x and, consequently, $D(x, y; \lambda_o) \not\equiv 0$ in x and y. Hence, indeed, in Theorem V the condition $D(x, y; \lambda_o) \not\equiv 0$ may be replaced by $D'(\lambda_o) \neq 0$.

We remark further that if $u(x) = D(x, y_o; \lambda_o)$ is a solution of the homogeneous integral equation (26), then Cu (C, an arbitrary constant) is also a solution. Thus, there are an infinitude of solutions which differ only by a constant factor. We shall later show that there are no other solutions. This is in analogy with the finite system of linear equations

$$u_i - \lambda h \sum_{j=1}^{n} K_{ij}u_j = 0 \qquad (i = 1, \ldots, n)$$

with determinant Δ. If $\Delta = 0$, while not all of the first minors Δ_{ik} vanish, then these n equations determine uniquely the ratios of the $u_1, \ldots, u_n$, that is, $u_j = C_j u_n$

[1] GOURSAT-HEDRICK, "Mathematical Analysis", vol. 1, §174.

$(j = 1, \ldots, n; C_n = 1)$. Now, $\Delta = 0$ corresponds to $D(\lambda) = 0$, while *not all* Δ_{ik} *vanish* corresponds to $D(x, y; \lambda_o) \not\equiv 0$.

21. Solution of the Homogeneous Integral Equation When D(λ) = 0.—It remains to consider the case

$$D(\lambda_o) = 0, \; D(x, y; \lambda_o) \equiv 0.$$

which corresponds in the linear system to the case where Δ and all its first minors are zero, and, where it becomes necessary, to consider the minors of higher order of Δ. Accordingly, we have to consider, in the treatment of the integral equation, the limits of these higher minors of Δ as h approaches zero.

In what follows let us use with Heywood-Frechet ("L'-Équation De Fredholm," page 53) the notation

$$K\begin{pmatrix} s_1, s_2, & \ldots & , s_n \\ t_1, t_2, & \ldots & , t_n \end{pmatrix} \equiv \begin{vmatrix} K(s_1, t_1) & \ldots & K(s_1, t_n) \\ \cdot & \cdot\cdot\cdot & \cdot \\ \cdot & \cdot\cdot\cdot & \cdot \\ K(s_n, t_1) & \ldots & K(s_n, t_n) \end{vmatrix}.$$

(*a*) *Definition of the pth Minor of* $D(\lambda)$.—Let

$$(28) \quad B_n\begin{pmatrix} x_1, & \ldots & , x_p \\ y_1, & \ldots & , y_p \end{pmatrix} = \int_a^b \ldots \int_a^b K\begin{pmatrix} x_1, & \ldots & , x_p; \; t_1, & \ldots & , t_n \\ y_1, & \ldots & , y_p, \; t_1, & \ldots & , t_n \end{pmatrix} dt_1 \ldots dt_n$$

with

$$(29) \quad B_o\begin{pmatrix} x_1, & \ldots & , x_p \\ y_1, & \ldots & , y_p \end{pmatrix} = K\begin{pmatrix} x_1, & \ldots & , x_p \\ y_1, & \ldots & , y_p \end{pmatrix}.$$

Then the *pth minor of* $D(\lambda)$ is defined by the infinite series

$$(30) \quad D\begin{pmatrix} x_1, & \ldots & , x_p & \lambda \\ y_1, & \ldots & , y_p & \end{pmatrix} = \sum_{n=0}^{\infty} (-1)^n \frac{\lambda^{n+p}}{n!} B_n\begin{pmatrix} x_1, & \ldots & , x_p \\ y_1, & \ldots & , y_p \end{pmatrix} \equiv D_p(x, y; \lambda),$$

which for $p = 1$ reduces to $D(x, y; \lambda)$.

By means of Hadamard's theorem we prove, exactly as before, the

Theorem VI.—*The infinite series for* $D\begin{pmatrix} x_1, & . & . & . & , x_p & \lambda \\ y_1, & . & . & . & , y_p & \end{pmatrix}$ *is absolutely and permanently convergent in* λ, *and uniformly convergent in* $x_1, \ldots, x_p, y_1, \ldots, y_p$ *for* $a \leq x_\alpha \leq b$, $a \leq y_\beta \leq b$ $(\alpha, \beta = 1, \ldots, p)$.

Corollary.—*When two of the* x's *become equal, say* $x_r = x_s$, *or when two of the* y's *become equal, say* $y_i = y_j$, *then* $D_p(x, y; \lambda)$ *vanishes.* For then in the integrand of $B_n\begin{pmatrix} x_1, & \ldots, & x_p \\ y_1, & \ldots, & y_p \end{pmatrix}$ two rows (columns) become equal. Therefore,

$$B_n\begin{pmatrix} x_1, & . & . & . & , x_p \\ y_1, & . & . & . & , y_p \end{pmatrix} = 0, \text{ and hence } D_p(x, y; \lambda) = 0.$$

In like manner, if in $D_p(x, y; \lambda)$ two of the x's or two of the y's are interchanged, $D_p(x, y; \lambda)$ changes sign.

b) Generalization of the Two Fundamental Relations.—Expand the determinant in $B_n\begin{pmatrix} x_1, & . & . & . & , x_p \\ y_1, & . & . & . & , y_p \end{pmatrix}$ according to the elements of the column y_β:

$$B_n\begin{pmatrix} x_1, & . & . & . & , x_p \\ y_1, & . & . & . & , y_p \end{pmatrix} = \int_a^b \cdots \int_a^b \Bigg\{ \sum_{\alpha=1}^{p} (-1)^{\alpha+\beta} \times$$

$$K(x_\alpha, y_\beta) K\begin{pmatrix} x_1, & . & . & . & , x_{\alpha-1}, x_{\alpha+1}, & . & . & . & , x_p, t_1, & . & . & . & , t_n \\ y_1, & . & . & . & , y_{\beta-1}, y_{\beta+1}, & . & . & . & , y_p, t_1, & . & . & . & , t_n \end{pmatrix}$$

$$+ \sum_{i=1}^{n} (-1)^{p+i+\beta} K(t_i, y_\beta) \times$$

$$K\begin{pmatrix} x_1, & . & . & . & , x_p, t_1, & . & . & . & , t_{i-1}, t_{i+1}, & . & . & . & , t_n \\ y_1, & . & . & . & , y_{\beta-1}, y_{\beta+1}, & . & . & . & , y_p, t_1, & . & . & . & , t_n \end{pmatrix} \Bigg\} dt_1 \ldots dt_n.$$

In the first sum, $K(x_\alpha, y_\beta)$ may be taken before the integral sign, and, according to (28), the sum becomes

$$\sum_{\alpha=1}^{p} (-1)^{\alpha+\beta} K(x_\alpha, y_\beta) B_n\begin{pmatrix} x_1, & . & . & . & , x_{\alpha-1}, x_{\alpha+1}, & . & . & . & , x_p \\ y_1, & . & . & . & , y_{\beta-1}, y_{\beta+1}, & . & . & . & , y_p \end{pmatrix}.$$

In the second sum make a change of notation.

In place of $\quad t_i, t_{i+1}, t_{i+2}, \ldots, t_n$
write $\quad t, t_i, t_{i+1}, \ldots, t_{n-1}$

The ith term of the second sum becomes

$$(-1)^{p+i+\beta}K(t, y_\beta) \times$$
$$K\begin{pmatrix} x_1, & \ldots & \ldots & \ldots, & x_p, & t_1, & \ldots & \ldots & \ldots & \ldots & \ldots, & t_{n-1} \\ y_1, & \ldots, & y_{\beta-1}, & y_{\beta+1}, & \ldots, & y_p, & t_1, & \ldots, & t_{i-1}, & t, & t_i, \ldots, & t_{n-1} \end{pmatrix}.$$

Bring the column t between the columns $y_{\beta-1}$, $y_{\beta+1}$ by $i + p - \beta - 1$ transpositions of columns. We get, then, for the ith term of the second sum

$$- K(t, y_\beta) \times$$
$$K\begin{pmatrix} x_1, & \ldots & x_{\beta-1}, & x_\beta, & x_{\beta+1}, & \ldots, & x_p, & t_1, & \ldots, & t_{n-1} \\ y_1, & \ldots, & y_{\beta-1}, & t, & y_{\beta+1}, & \ldots, & y_p, & t_1, & \ldots, & t_{n-1} \end{pmatrix}.$$

Hence, all terms in $\sum_{i=1}^{n}$ are equal and this sum may be written, if, moreover, we integrate first with respect to $t_1, \ldots, t_{n-1}$,

$$- n\int_a^b K(t, y_\beta)\left\{\int_a^b \ldots \int_a^b K\begin{pmatrix} x_1, & \ldots & \ldots & \ldots & \ldots, & x_p, & t_1, & \ldots, & t_{n-1} \\ y_1, & \ldots, & y_{\beta-1}, & t, & y_{\beta+1}, \ldots, & y_p, & t_1, & \ldots, & t_{n-1} \end{pmatrix} dt_1 \ldots dt_{n-1}\right\} dt,$$

which, according to (28), reduces to

$$- n\int_a^b K(t, y_\beta) \times B_{n-1}\begin{pmatrix} x_1, & \ldots, & x_{\beta-1}, & x_\beta, & x_{\beta+1}, & \ldots & x_p \\ y_1, & \ldots, & y_{\beta-1}, & t, & y_{\beta+1}, & \ldots, & y_p \end{pmatrix} dt.$$

Thus we arrive at the formula

$$(31) \qquad B_n\begin{pmatrix} x_1, & \ldots & , x_p \\ y_1, & \ldots & , y_p \end{pmatrix} =$$

$$\sum_{\alpha=1}^{p}(-1)^{\alpha+\beta}K(x_\alpha, y_\beta)B_n\begin{pmatrix} x_1, & \ldots & , x_{\alpha-1}, x_{\alpha+1}, & \ldots & , x_p \\ y_1, & \ldots & , y_{\beta-1}, y_{\beta+1}, & \ldots & , y_p \end{pmatrix}$$

$$- n\int_a^b K(t, y_\beta)B_{n-1}\begin{pmatrix} x_1, & \ldots & , x_{\alpha-1}, x_\alpha, x_{\alpha+1}, & \ldots & , x_p \\ y_1, & \ldots & , y_{\beta-1}, t, y_{\beta+1}, & \ldots & , y_p \end{pmatrix}dt.$$

In like manner, by expanding the integrand of $B_n\begin{pmatrix} x_1, & \ldots & , x_p \\ y_1, & \ldots & , y_p \end{pmatrix}$ according to the elements of the *row* x_α, we obtain

$$(32) \qquad B_n\begin{pmatrix} x_1, & \ldots & , x_p \\ y_1, & \ldots & , y_p \end{pmatrix} =$$

$$\sum_{\beta=1}^{p}(-1)^{\alpha+\beta}K(x_\alpha, y_\beta)B_n\begin{pmatrix} x_1, & \ldots & , x_{\alpha-1}, x_{\alpha+1}, & \ldots & , x_p \\ y_1, & \ldots & , y_{\beta-1}, y_{\beta+1}, & \ldots & , y_p \end{pmatrix}$$

$$- n\int_a^b K(x_\alpha, t)B_{n-1}\begin{pmatrix} x_1, & \ldots & , x_{\alpha-1}, t, x_{\alpha+1}, & \ldots & , x_p \\ y_1, & \ldots & , y_{\beta-1}, y_\beta, y_{\beta+1}, & \ldots & , y_p \end{pmatrix}dt.$$

If now we multiply both sides of (31) and (32) by $(-1)^n \dfrac{\lambda^{n+p}}{n!}$ and sum with respect to n from $n = 0$ to $n = \infty$, we obtain, on account of the definition given in (30), the following double relation, which is a *generalization of Fredholm's two fundamental relations* (10) and (11).

$$(33) \qquad D\begin{pmatrix} x_1, & \ldots & , x_p \\ y_1, & \ldots & , y_p \end{pmatrix} \lambda =$$

$$\sum_{\alpha=1}^{p}(-1)^{\alpha+\beta}\lambda K(x_\alpha, y_\beta)D\begin{pmatrix} x_1, & \ldots & , x_{\alpha-1}, x_{\alpha+1}, & \ldots & , x_p \\ y_1, & \ldots & , y_{\beta-1}, y_{\beta+1}, & \ldots & , y_p \end{pmatrix} \lambda$$

$$+ \lambda\int_a^b K(t, y_\beta)D\begin{pmatrix} x_1, & \ldots & , x_{\beta-1}, x_\beta, x_{\beta+1}, & \ldots & , x_p \\ y_1, & \ldots & , y_{\beta-1}, t, y_{\beta+1}, & \ldots & , y_p \end{pmatrix} \lambda dt$$

$$(34) \quad = \sum_{\beta=1}^{p} (-1)^{\alpha+\beta} \lambda K(x_\alpha, y_\beta)$$

$$D\begin{pmatrix} x_1, & \ldots, & x_{\alpha-1}, x_{\alpha+1}, & \ldots, & x_p \\ y_1, & \ldots, & y_{\beta-1}, y_{\beta+1}, & \ldots, & y_p \end{pmatrix} \lambda$$

$$+ \lambda \int_a^b K(x_\alpha, t) D\begin{pmatrix} x_1, & \ldots, & x_{\alpha-1}, t, x_{\alpha+1}, & \ldots, & x_p \\ y_1, & \ldots, & y_{\alpha-1}, y_\alpha, y_{\alpha+1}, & \ldots, & y_p \end{pmatrix} \lambda \, dt.$$

c) Relation between $D^{(p)}(\lambda)$ *and* $D\begin{pmatrix} x_1, & \ldots, & x_p \\ y_1, & \ldots, & y_p \end{pmatrix} \lambda$.—

The relation (27) between $D'(\lambda)$ and $D(x, x; \lambda)$ generalizes as follows:

$$(35) \quad \int_a^b \ldots \int_a^b D\begin{pmatrix} x_1, & \ldots, & x_p \\ x_1, & \ldots, & x_p \end{pmatrix} \lambda \, dx_1 \ldots dx_p = (-1)^p \lambda^p \frac{d^p D(\lambda)}{d\lambda^p}.$$

Proof.—From the series expression for $D(\lambda)$:

$$D(\lambda) = 1 + \sum_{n=1}^{\infty} (-1)^n \frac{\lambda^n}{n!} A_n,$$

we get

$$D^{(p)}(\lambda) = \sum_{n=p}^{\infty} (-1)^n \frac{\lambda^{n-p}}{(n-p)!} A_n,$$

which, after the change in notation $n - p = n'$ and a final dropping of the prime ($'$), may be written

$$D^{(p)}(\lambda) = \sum_{n=0}^{\infty} (-1)^{n+p} \frac{\lambda^n}{n!} A_{n+p}.$$

But by (15)

$$A_{n+p} = \int_a^b \ldots \int_a^b K\begin{pmatrix} t_1, & \ldots, & t_{n+p} \\ t_1, & \ldots, & t_{n+p} \end{pmatrix} dt_1 \ldots dt_{n+p}.$$

If now

in place of $\quad t_1, \ldots, t_p, t_{p+1}, \ldots, t_{p+n}$

we write $\quad x_1, \ldots, x_p, t_1, \ldots, t_n$

and then change the order of integration, as we may, so that we integrate first with respect to $t_1, \ldots, t_n$, we obtain

$$A_{n+p} = \int_a^b \cdots \int_a^b \left\{ \int_a^b \cdots \int_a^b K\begin{pmatrix} x_1, & \ldots, & x_p, & t_1, & \ldots, & t_n \\ x_1, & \ldots, & x_p, & t_1, & \ldots, & t_n \end{pmatrix} dt_1 \ldots dt_n \right\} dx_1 \ldots dx_p,$$

which, on account of (28), becomes

$$A_{n+p} = \int_a^b \cdots \int_a^b B_n\begin{pmatrix} x_1, & \ldots, & x_p \\ x_1, & \ldots, & x_p \end{pmatrix} dx_1 \ldots dx_p.$$

Multiply both sides of this equality by $(-1)^n \dfrac{\lambda^{n+p}}{n!}$ and sum with respect to n from $n = 0$ to $n = \infty$. We obtain

$$(-1)^p \lambda^p \frac{d^pD(\lambda)}{d\lambda^p} = \sum_{n=0}^{\infty} \int_a^b \cdots \int_a^b (-1)^n \frac{\lambda^{n+p}}{n!} B_n\begin{pmatrix} x_1, & \ldots, & x_p \\ x_1, & \ldots, & x_p \end{pmatrix} dx_1 \ldots dx_p.$$

It is permissible here to put the summation under the multiple integral sign. If we then make use of the equation of definition (30), we obtain

$$(-1)^p \lambda^p \frac{d^pD(\lambda)}{d\lambda^p} = \int_a^b \cdots \int_a^b D\begin{pmatrix} x_1, & \ldots, & x_p \\ x_1, & \ldots, & x_p \end{pmatrix} \lambda \Bigg) dx_1 \ldots dx_p,$$

which establishes the equality (35).

We make use of this result to prove that not all of the Fredholm minors vanish. Let λ_o be a root of $D(\lambda) = 0$. Then certainly $\lambda_o \neq 0$ for $D(0) = 1$. Furthermore, λ_o is a root of $D(\lambda)$ of finite *multiplicity* r $(r \geq 1)$, defined by

$$D(\lambda_o) = 0, D'(\lambda_o) = 0, \ldots, D^{(r-1)}(\lambda_o) = 0, D^{(r)}(\lambda_o) \neq 0.$$

The multiplicity r of the root must be finite, otherwise

$$D(\lambda) \equiv 0.$$

In (35) put $\lambda = \lambda_o$ and $p = r$, then the right member by our hypothesis does not vanish. Therefore, the left member does not vanish. Hence, since $\lambda_o \neq 0$,

$$D\begin{pmatrix} x_1, & . & . & . & , x_r \\ x_1, & . & . & . & , x_r \end{pmatrix} \lambda_o \Big) \not\equiv 0 \text{ in } x_1, \ . \ . \ . \ , x_r$$

and, consequently,

$$D\begin{pmatrix} x_1, & . & . & . & , x_r & \lambda_o \\ y_1, & . & . & . & , y_r & \end{pmatrix} \not\equiv 0 \text{ in } x_1, \ . \ . \ . \ , x_r, y_1, \ . \ . \ . \ , y_r.$$

Hence, in the series

$$D(\lambda_o) = 0,\ D\begin{pmatrix} x & \lambda_o \\ y & \end{pmatrix},\ D\begin{pmatrix} x_1, x_2 & \lambda_o \\ y_1, y_2 & \end{pmatrix},\ D\begin{pmatrix} x_1, x_2, x_3 & \lambda_o \\ y_1, y_2, y_3 & \end{pmatrix},\ . \ . \ .$$

we must come to a number $q \leqq r$, called the *index of* λ_o, such that

$$D(\lambda_o) = 0,\ D(x, y; \lambda_o) \equiv 0,\ . \ . \ . \ ,$$
$$D\begin{pmatrix} x_1, & . & . & . & , x_{q-1} & \lambda_o \\ y_1, & . & . & . & , y_{q-1} & \end{pmatrix} \equiv 0,\ D\begin{pmatrix} x_1, & . & . & . & , x_q & \lambda_o \\ y_1, & . & . & . & , y_q & \end{pmatrix} \not\equiv 0.$$

That is, there exists a particular set of values x_1', . . . , x_q', y_1', . . . , y_q' for the variables x_1, . . . , x_q, y_1, . . . , y_q such that we have the following numerical inequality:

$$D\begin{pmatrix} x_1', & . & . & . & , x_q' & \lambda_o \\ y_1', & . & . & . & , y_q' & \end{pmatrix} \neq 0.$$

Incidentally, we have proved the

Theorem VII.—*The index q of a root* λ_o *of* $D(\lambda)$ *is at most equal to the multiplicity r of* λ_o*:* $q \leqq r$.

d) The q Independent Solutions of the Homogeneous Equation.—Let λ_o be a root of $D(\lambda) = 0$ of index q, so that

$$D(\lambda_o) = 0,\ D(x, y; \lambda_o) \equiv 0,\ . \ . \ . \ ,$$
$$D\begin{pmatrix} x_1, & . & . & . & , x_{q-1} & \lambda_o \\ y_1, & . & . & . & , y_{q-1} & \end{pmatrix} \equiv 0$$

but
$$D\begin{pmatrix} x_1, & . & . & . & , x_q & \lambda_o \\ y_1, & . & . & . & , y_q & \end{pmatrix} \not\equiv 0.$$

Write down the second of Fredholm's generalized fundamental relations (34) for $\lambda = \lambda_o$, $p = q$

$$x_1 = x_1', \quad . \quad . \quad . \quad , x_{\alpha-1} = x'_{\alpha-1}, \; x_\alpha = x, \; x_{\alpha+1} = x'_{\alpha+1}, \quad . \quad . \quad . \quad , x_q = x_q'$$

$$y_1 = y_1', \quad . \quad . \quad . \quad , y_{\beta-1} = y'_{\beta-1}, \; y_\beta = y_\beta', \; y_{\beta+1} = y'_{\beta+1}, \quad . \quad . \quad . \quad , y_q = y_q'$$

Then

$$D\begin{pmatrix} x_1', & . \; . \; . \; , x'_{\alpha-1}, & x, & x'_{\alpha+1}, & . \; . \; . \; , x_q' & \\ y_1', & . \; . \; . \; , y'_{\beta-1}, & y_\beta', & y'_{\beta+1}, & . \; . \; . \; , y_q' & \lambda_o \end{pmatrix} =$$

$$\lambda_o \int_a^b K(x, t) D\begin{pmatrix} x_1', & . \; . \; . \; , x'_{\alpha-1}, & t, & x'_{\alpha+1}, & . \; . \; . \; , x_q' & \\ y_1', & . \; . \; . \; , y'_{\beta-1}, & y_\beta', & y'_{\beta+1}, & . \; . \; . \; , y_q' & \lambda_o \end{pmatrix} dt,$$

since by hypothesis

$$D\begin{pmatrix} x_1, & . \; . \; . \; , x_{\alpha-1}, & x_{\alpha+1}, & . \; . \; . \; , x_q & \\ y_1, & . \; . \; . \; , y_{\beta-1}, & y_{\beta+1}, & . \; . \; . \; , y_q & \lambda_o \end{pmatrix} \equiv 0.$$

Hence, if we divide by $D\begin{pmatrix} x_1', & . \; . \; . \; , x_q' & \\ y_1', & . \; . \; . \; , y_q' & \lambda_o \end{pmatrix}$

and put

$$(36) \quad D\begin{pmatrix} x_1', & . \; . \; . \; , x'_{\alpha-1}, & x \;, & x'_{\alpha+1}, & . \; . \; . \; , x_q' & \\ y_1', & . \; . \; . \; , y'_{\beta-1}, & y'_\beta, & y'_{\beta+1}, & . \; . \; . \; , y_q' & \lambda_o \end{pmatrix} = \varphi_\alpha(x, \lambda_o) D\begin{pmatrix} x_1', & . \; . \; . \; , x_q' & \\ y_1', & . \; . \; . \; , y_q' & \lambda_o \end{pmatrix}$$

we have

$$\varphi_\alpha(x, \lambda_o) = \lambda_o \int_a^b K(x, t) \varphi_\alpha(t, \lambda_o) dt,$$

which expresses that the q functions

$$\varphi_1(x, \lambda_o), \; \varphi_2(x, \lambda_o), \quad . \quad . \quad . \quad , \varphi_q(x, \lambda_o)$$

are solutions of the homogeneous equation (26). These solutions are continuous and

$$(37) \qquad \varphi_\alpha(x_\beta', \lambda_o) = \begin{cases} 1, \; \beta = \alpha \\ 0, \; \beta \neq \alpha \end{cases}.$$

from (30) and (28). Furthermore, these functions are linearly independent. That is, if a relation of the form

$$C_1\varphi_1(x) + C_2\varphi_2(x) + \cdots + C_q\varphi_q(x) \equiv 0$$

exists, where $C_1, \ldots, C_q$ are constants, then we must have $C_1 = C_2 = \cdots = C_q = 0$. For, from (37), if $x = x_\alpha'$, then $C_\alpha = 0$.

From the homogeneity of (26) it follows that

$$u(x) = C_1\varphi_1(x) + \cdots + C_q\varphi_q(x) \tag{38}$$

is again a solution of (26) for arbitrary values of $C_1, \ldots, C_q$. We thus have a q-fold infinitude of solutions.

e) Completeness Proof.—It remains to show that every solution of (26) can be put in the form (38).

If $v(x)$ is any solution of (26), then

$$v(x) = \lambda_o \int_a^b K(x, t)v(t)dt.$$

Whence

$$0 \equiv \int_a^b H(x, t)\left\{v(t) - \lambda_o \int_a^b K(t, s)v(s)ds\right\}dt,$$

where $H(x, t)$ is any continuous function. On subtracting the second equation from the first, we obtain

$$v(x) = \lambda_o \int_a^b N(x, t)v(t)dt \tag{39}$$

where

$$\lambda_o N(x, t) = \lambda_o K(x, t) - \left\{H(x, t) - \lambda_o \int_a^b H(x, s)K(s, t)ds\right\}\cdot$$

Now apply (33) with $p = q + 1$, $x_{q+1} = x$, $y_{q+1} = y$ and notice that a transposition of two x's or two y's changes the sign of D. Then

$$D\begin{pmatrix} x, x_1, & \ldots, x_q \\ y, y_1, & \ldots, y_q \end{pmatrix} \lambda_o = \lambda K(x, y) D\begin{pmatrix} x_1, & \ldots, x_q \\ y_1, & \ldots, y_q \end{pmatrix}\lambda$$
$$-\sum_{\alpha=1}^{q} \lambda K(x_\alpha, y) D\begin{pmatrix} x_1, & \ldots, x_{\alpha-1}, x, x_{\alpha+1}, & \ldots, x_q \\ y_1, & \ldots, y_{\beta-1}, y_\beta, y_{\beta+1}, & \ldots, y_q \end{pmatrix}\lambda$$
$$+\lambda \int_a^b K(t, y) D\begin{pmatrix} x, x_1, & \ldots, x_q \\ t, y_1, & \ldots, y_q \end{pmatrix}\lambda\, dt.$$

Put $x_1 = x_1', \ldots, x_q = x_q', y_1 = y_1', \ldots, y_q = y_q'$, $y = t$, $\lambda = \lambda_o$, divide by

$$D\begin{pmatrix} x_1', & \ldots, & x_q' \\ y_1', & \ldots, & y_q' \end{pmatrix}\lambda_o\Big)$$

and put

$$(40) \qquad H(x, y) = \frac{D\begin{pmatrix} x, & x_1', & \ldots, & x_q' \\ y, & y_1', & \ldots, & y_q' \end{pmatrix}\lambda_o\Big)}{D\begin{pmatrix} x_1', & \ldots, & x_q' \\ y_1', & \ldots, & y_q' \end{pmatrix}\lambda_o\Big)},$$

then we obtain

$$(41) \qquad \sum_{\alpha=1}^{q} \lambda_o K(x_\alpha', t)\varphi_\alpha(x) = \lambda_o K(x, t) - H(x, t) + \lambda_o \int_a^b H(x, s)K(s, t)ds.$$

The equation (39) can now be written

$$v(x) = \lambda_o \sum_{\alpha=1}^{q} \varphi_\alpha(x) \int_a^b K(x_\alpha', t)v(t)dt.$$

This shows that $v(x)$ can be written in the form (38) by taking for the constants C_α the values

$$C_\alpha = \lambda_o \int_a^b K(x_\alpha', t)v(t)dt.$$

Thus we obtain *Fredholm's second fundamental theorem:*

Theorem VIII.—*If* $\lambda = \lambda_o$ *is a root of* $D(\lambda) = 0$ *of order* q, *then the homogeneous integral equation*

$$(26) \qquad u(x) = \lambda_o \int_a^b K(x, t)u(t)dt$$

has q *linearly independent solutions in terms of which every other solution is expressible linearly and homogeneously. Such a system of* q *independent solutions is given by*

$$\varphi_\alpha(x) = \frac{D\begin{pmatrix} x_1', & \ldots, & x'_{\alpha-1}, & x, & x'_{\alpha+1}, & \ldots, & x_q' \\ y_1', & \ldots, & y'_{\beta-1}, & y_\beta', & y'_{\beta+1}, & \ldots, & y_q' \end{pmatrix}\lambda_o\Big)}{D\begin{pmatrix} x_1', & \ldots, & x_q' \\ y_1', & \ldots, & y_q' \end{pmatrix}\lambda_o\Big)}$$

$$(\alpha = 1, \ldots, q).$$

22. Characteristic Constant. Fundamental Functions. Definitions.—If $D(\lambda)$ is the Fredholm's determinant for the kernel $K(x, t)$ and $D(\lambda_o) = 0$, then λ_o is said to be a *characteristic constant* of the kernel $K(x, t)$. Further, if $\varphi(x)$ is continuous and not identically zero on the interval (ab) and

$$\varphi(x) = \lambda_o \int_a^b K(x, t)\varphi(t)dt,$$

then $\varphi(x)$ is called a *fundamental function* of the kernel $K(x, t)$, belonging to the characteristic constant λ_o.

$\varphi_1(x), \ldots, \varphi_q(x)$ form a *complete system* of fundamental functions of the kernel $K(x, t)$, belonging to λ_o, if every other solution is expressible linearly in terms of these q solutions. Thus, if $\psi_1, \ldots, \psi_q$ are any other q solutions, we must have

$$\begin{aligned} \psi_1 &= C_{11}\varphi_1 + \ldots + C_{1q}\varphi_q \\ &\cdots\cdots\cdots\cdots \\ \psi_q &= C_{q1}\varphi_1 + \ldots + C_{qq}\varphi_q \end{aligned}$$

and if

$$\begin{vmatrix} C_{11} & \ldots & C_{1q} \\ \cdot & \cdot & \cdot \\ \cdot & \cdot & \cdot \\ C_{q1} & \ldots & C_{qq} \end{vmatrix} \neq 0$$

$\psi_1, \ldots, \psi_q$ form again a complete system of fundamental functions belonging to λ_o.

23. The Associated Homogeneous Integral Equation.—Preparatory to the discussion of the non-homogeneous integral equation, for $D(\lambda) = 0$, we will discuss the homogeneous integral equation

$$v(x) = \lambda_o \int_a^b K(t, x)v(t)dt, \tag{42}$$

which is called the integral equation *associated* with the integral equation

$$u(x) = \lambda_o \int_a^b K(x, t)u(t)dt. \tag{26}$$

Notice that the kernel

$$\overline{K}(x, t) \equiv K(t, x)$$

of the associated equation is derived from the original kernel $K(x, t)$ by interchanging the arguments x and t. There exist important relations between the solutions of the two equations (42) and (26). To obtain them, we first compute the Fredholm determinant and Fredholm minors for the kernel $\overline{K}(x, t)$, which we indicate by the corresponding dashed notation.

a) Fredholm's Determinant for the Associated Kernel.—

$$D(\lambda) = 1 + \sum_{n=1}^{\infty} (-1)^n \frac{\lambda^n}{n!} A_n$$

where

$$A_n = \int_a^b \cdots \int_a^b \begin{vmatrix} K(t_1, t_1) & \cdots & K(t_1, t_n) \\ \cdot & \cdots & \cdot \\ \cdot & \cdots & \cdot \\ \cdot & \cdots & \cdot \\ K(t_n, t_1) & \cdots & K(t_n, t_n) \end{vmatrix} dt_1 \cdots dt_n.$$

Then

$$\overline{D}(\lambda) = 1 + \sum_{n=1}^{\infty} (-1)^n \frac{\lambda^n}{n!} \overline{A}_n$$

where

$$\overline{A}_n = \int_a^b \cdots \int_a^b \begin{vmatrix} (\overline{K}t_1, t_1) & \cdots & \overline{K}(t_1, t_n) \\ \cdot & \cdots & \cdot \\ \cdot & \cdots & \cdot \\ \cdot & \cdots & \cdot \\ \overline{K}(t_n, t_1) & \cdots & \overline{K}(t_n, t_n) \end{vmatrix} dt_1 \cdots dt_n.$$

In this expression for $\overline{A}_n$ put $\overline{K}(x, t) = K(t, x)$. Then

$$\overline{A}_n = \int_a^b \cdots \int_a^b \begin{vmatrix} K(t_1, t_1) & \cdots & K(t_n, t_1) \\ \cdot & \cdot & \cdot \\ \cdot & \cdot & \cdot \\ \cdot & \cdot & \cdot \\ K(t_1, t_n) & \cdots & K(t_n, t_n) \end{vmatrix} dt_1 \ldots dt_n.$$

The determinant which appears in the integrand of $\overline{A}_n$ is the same as that which appears in the integrand of the expression for A_n, with the exception that the rows and columns are interchanged. But this interchange leaves the value of the determinant unaltered. Therefore $\overline{A}_n = A_n$, and hence

$$\overline{D}(\lambda) \equiv D(\lambda). \tag{43}$$

Hence also we conclude that $K(t, x)$ and $K(x, t)$ have the same characteristic constants.

b) Fredholm's Minors for the Associated Kernel.—We have from (30)

$$D\begin{pmatrix} x_1, & \ldots & , x_p \\ y_1, & \ldots & , y_p \end{pmatrix}\lambda = \sum_{n=0}^{\infty} (-1)^n \frac{\lambda^{n+p}}{n!} B_n\begin{pmatrix} x_1, & \ldots & , x_p \\ y_1, & \ldots & , y_p \end{pmatrix}$$

where the B_n are given by (28). Then

$$\overline{B}_n\begin{pmatrix} x_1, & \ldots & , x_p \\ y_1, & \ldots & , y_p \end{pmatrix} = \int_a^b \cdots \int_a^b$$

$$\begin{vmatrix} \overline{K}(x_1, y_1) & \cdots & \overline{K}(x_1, y_p) & \overline{K}(x_1, t_1) & \cdots & \overline{K}(x_1, t_n) \\ \cdot & \cdot & \cdot & \cdot & \cdot & \cdot \\ \cdot & \cdot & \cdot & \cdot & \cdot & \cdot \\ \overline{K}(x_p, y_1) & \cdots & \overline{K}(x_p, y_p) & \overline{K}(x_p, t_1) & \cdots & \overline{K}(x_p, t_n) \\ \overline{K}(t_1, y_1) & \cdots & \overline{K}(t_1, y_p) & \overline{K}(t_1, t_1) & \cdots & \overline{K}(t_1, t_n) \\ \cdot & \cdot & \cdot & \cdot & \cdot & \cdot \\ \cdot & \cdot & \cdot & \cdot & \cdot & \cdot \\ \overline{K}(t_n, y_1) & \cdots & \overline{K}(t_n, y_p) & \overline{K}(t_n, t_1) & \cdots & \overline{K}(t_n, t_n) \end{vmatrix}$$

$$dt_1 \ldots dt_n.$$

But $\overline{K}(x, t) = K(t, x)$, then

$$\overline{B}_n\begin{pmatrix} x_1, & . & . & . & , x_p \\ y_1, & . & . & . & , y_p \end{pmatrix} = \int_a^b \cdots \int_a^b$$

$$\begin{vmatrix} K(y_1, x_1) & . & . & . & K(y_p, x_1) & K(t_1, x_1) & . & . & . & K(t_n, x_1) \\ . & . & . & . & . & . & . & . & . & . \\ . & . & . & . & . & . & . & . & . & . \\ K(y_1, x_p) & . & . & . & K(y_p, x_p) & K(t_1, x_p) & . & . & . & K(t_n, x_p) \\ K(y_1, t_1) & . & . & . & K(y_p, t_1) & K(t_1, t_1) & . & . & . & K(t_n, t_1) \\ . & . & . & . & . & . & . & . & . & . \\ . & . & . & . & . & . & . & . & . & . \\ K(y_1, t_n) & . & . & . & K(y_p, t_n) & K(t_1, t_n) & . & . & . & K(t_n, t_n) \end{vmatrix} dt_1 \ . \ . \ . \ dt_n.$$

An interchange of rows and columns in the determinant of the integrand does not change the value of the determinant. Hence

$$\overline{B}_n\begin{pmatrix} x_1, & . & . & . & , x_p \\ y_1, & . & . & . & , y_p \end{pmatrix} = \int_a^b \cdots \int_a^b K\begin{pmatrix} y_1, & . & . & . & , y_p, t_1, & . & . & . & , t_n \\ x_1, & . & . & . & , x_p, t_1, & . & . & . & , t_n \end{pmatrix} dt_1 \ . \ . \ . \ dt_n.$$

Then, by means of the equation (28) defining B_n, we see that

$$\overline{B}_n\begin{pmatrix} x_1, & . & . & . & , x_p \\ y_1, & . & . & . & , y_p \end{pmatrix} = B_n\begin{pmatrix} y_1, & . & . & . & , y_p \\ x_1, & . & . & . & , x_p \end{pmatrix}$$

and, therefore,

$$(44) \qquad \overline{D}\begin{pmatrix} x_1, & . & . & . & , x_p \\ y_1, & . & . & . & , y_p \end{pmatrix} \lambda \Bigg) \equiv D\begin{pmatrix} y_1, & . & . & . & , y_p \\ x_1, & . & . & . & , x_p \end{pmatrix} \lambda \Bigg).$$

Since λ_o is a characteristic constant of $K(x, t)$ of index q, we have

$$D\begin{pmatrix} x_1, & . & . & . & , x_p \\ y_1, & . & . & . & , y_p \end{pmatrix} \lambda_o \Bigg) \equiv 0 \qquad (p = 1, \ . \ . \ . \ , q - 1)$$

while

$$D\begin{pmatrix} x_1', & . & . & . & , x_q' \\ y_1', & . & . & . & , y_q' \end{pmatrix} \lambda_o \Bigg) \neq 0.$$

Hence, by (44)

$$\bar{D}\begin{pmatrix} x_1, & \cdots & , x_p \\ y_1, & \cdots & , y_p \end{pmatrix} = D\begin{pmatrix} y_1, & \cdots & , y_p \\ x_1, & \cdots & , x_p \end{pmatrix} \lambda_o \Bigg) \equiv 0$$

in $x_1, \ldots, x_p, y_1, \ldots y_p$ for $p = 1, \ldots, q - 1$.

Further, if we put

$$\bar{x}_\alpha' = y_\alpha', \quad \bar{y}_\alpha' = x_\alpha', \tag{45}$$

we have

$$\bar{D}\begin{pmatrix} \bar{x}_1', & \cdots & , \bar{x}_q' \\ \bar{y}_1', & \cdots & , \bar{y}_q' \end{pmatrix} \lambda_o \Bigg) = D\begin{pmatrix} \bar{y}_1', & \cdots & , \bar{y}_q' \\ \bar{x}_1', & \cdots & , \bar{x}_q' \end{pmatrix} \lambda_o \Bigg)$$
$$= D\begin{pmatrix} x_1', & \cdots & , x_q' \\ y_1', & \cdots & , y_q' \end{pmatrix} \lambda_o \Bigg) \neq 0.$$

Hence, by definition, the index $\bar{q}$ of λ_o as a characteristic constant of $K(t, x)$ is q. We state this result in the

Theorem IX.—*If λ_o is a characteristic constant of $K(x, t)$ of index q, then λ_o is a characteristic constant of $K(t, x)$ of the same index:*

$$\bar{q} = q.$$

c) *The Fundamental Functions of the Associated Equation.* Apply, now, Theorem VIII to equation (42) and we find that it has q linearly independent solutions. A fundamental system of such solutions is given by

$$\bar{\varphi}_\alpha(x) = \frac{\bar{D}\begin{pmatrix} \bar{x}_1', & \cdots & , \bar{x}'_{\alpha-1}, & x, & \bar{x}'_{\alpha+1}, & \cdots & , \bar{x}_q' \\ \bar{y}_1', & \cdots & , \bar{y}'_{\alpha-1}, & \bar{y}_\alpha', & \bar{y}'_{\alpha+1}, & \cdots & , \bar{y}_q' \end{pmatrix} \lambda_o \Bigg)}{\bar{D}\begin{pmatrix} \bar{x}_1', & \cdots & , \bar{x}_q' \\ \bar{y}_1', & \cdots & , \bar{y}_q' \end{pmatrix} \lambda_o \Bigg)}$$
$$= \frac{D\begin{pmatrix} \bar{y}_1', & \cdots & , \bar{y}'_{\alpha-1}, & \bar{y}_\alpha', & \bar{y}'_{\alpha+1}, & \cdots & , \bar{y}_q' \\ \bar{x}_1', & \cdots & , \bar{x}'_{\alpha-1}, & x, & \bar{x}'_{\alpha+1}, & \cdots & , \bar{x}_q' \end{pmatrix} \lambda_o \Bigg)}{D\begin{pmatrix} \bar{y}_1', & \cdots & , \bar{y}_q' \\ \bar{x}_1', & \cdots & , \bar{x}_q' \end{pmatrix} \lambda_o \Bigg)}.$$

If, now, we introduce the change of notation (45), we find

$$\bar{\varphi}_\alpha(x) = \frac{D\begin{pmatrix} x_1', & \dots, & x'_{\alpha-1}, x_\alpha', x'_{\alpha+1}, & \dots, & x_q' \\ y_1', & \dots, & y'_{\alpha-1}, x\ , y'_{\alpha+1}, & \dots, & y_q' \end{pmatrix}\lambda_o}{D\begin{pmatrix} x_1', & \dots, & x_q' \\ y_1', & \dots, & y_q' \end{pmatrix}\lambda_o}$$

$$(\varphi = 1, \dots, q).$$

The most general solution of (42) is now

$$v(x) = C_1\bar{\varphi}_1(x) + C_2\bar{\varphi}_2(x) + \dots + C_q\bar{\varphi}_q(x).$$

d) The Function $H(x, y)$ for the Associated Kernel.—From the definition of $H(x, y)$ given earlier, we have

$$\bar{H}(x, y) = \frac{\bar{D}\begin{pmatrix} x, \bar{x}_1', & \dots, & \bar{x}_q' \\ y, \bar{y}_1', & \dots, & \bar{y}_q' \end{pmatrix}\lambda_o}{\bar{D}\begin{pmatrix} \bar{x}_1', & \dots, & \bar{x}_q' \\ \bar{y}_1', & \dots, & \bar{y}_q' \end{pmatrix}\lambda_o}.$$

Apply (44) and make the change of notation (45). Then

$$\text{(46)} \qquad \bar{H}(x, y) = \frac{D\begin{pmatrix} y, y_1', & \dots, & y_q' \\ x, x_1', & \dots, & x_q' \end{pmatrix}\lambda_o}{D\begin{pmatrix} y_1', & \dots, & y_q' \\ x_1', & \dots, & x_q' \end{pmatrix}\lambda_o} = H(y, x).$$

If, now, we take account of (46) and make the change of notation (45), then the relation (41) written for the kernel $K(x, t) = K(t, x)$ becomes

$$\text{(47)} \quad \sum_{\alpha=1}^{q} \lambda_o K(t, y_\alpha')\bar{\varphi}_\alpha(x) = \lambda_o K(t, x) - H(t, x) + \lambda_o \int_a^b H(s, x)K(t, s)ds.$$

e) The Orthogonality Theorem.—

Theorem X.—*If λ_o and λ_1 are two distinct characteristic constants of $K(x, t)$, $\varphi_o(x)$ is a fundamental function of*

$K(x, t)$ for λ_o, and $\bar{\varphi}_1(x)$ is a fundamental function of $\bar{K}(x, t)$ for λ_1, that is

$$(48)\quad \varphi_o(x) = \lambda_o \int_a^b K(x, t)\varphi_o(t)dt$$

$$(49)\quad \bar{\varphi}_1(x) = \lambda_1 \int_a^b \bar{K}(x, t)\bar{\varphi}_1(t)dt = \lambda_1 \int_a^b K(t, x)\bar{\varphi}_1(t)dt$$

then

$$(50)\qquad \int_a^b \varphi_o(x)\bar{\varphi}_1(x) = 0.$$

Proof.—From (48) and (49) we obtain

$$(\lambda_o - \lambda_1)\int_a^b \varphi_o(x)\bar{\varphi}_1(x)dx =$$

$$\lambda_o\lambda_1 \int_a^b \int_a^b \varphi_o(x)K(t, x)\bar{\varphi}_1(t)\, dt\, dx$$

$$- \lambda_o\lambda_1 \int_a^b \int_a^b \bar{\varphi}_1(x)K(x, t)\varphi_o(t)\, dt\, dx.$$

We see that the two integrals on the right are equal if in the last integral we write t and x in place of x and t. Then, since by hypothesis $\lambda_o \neq \lambda_1$, we must have (50).

Definition.—Two continuous functions $g(x)$, $h(x)$, for which

$$\int_a^b g(x)h(x)dx = 0,$$

are said to be *orthogonal* to each other.

Hence the above result may be stated as follows: $\varphi_o(x)$ and $\bar{\varphi}_1(x)$ are orthogonal to each other.

24. The Non-homogeneous Integral Equation When $D(\lambda) = 0$.—With the aid of the results established in the last article we can discuss completely the solution of the non-homogeneous integral equation

$$(51)\qquad u(x) = f(x) + \lambda_o \int_a^b K(x, t)u(t)dt$$

when $D(\lambda_o) = 0$ and λ_o is of index q.

The finite system of linear equations

$$u_i - \lambda h \sum_{j=1}^{n} K_{ij} u_j = f_i \qquad (i = 1, \ldots, n),$$

of which the equation (51) may be considered as a limit, has, in general, for $\Delta = 0$, no finite solution. If, however, certain conditions on the f_i are satisfied, the system has an infinitude of solutions. In analogy we will find that (51), for $D(\lambda_o) = 0$, has, in general, no solutions. If, however, $f(x)$ satisfies certain conditions, then (51) has an infinitude of solutions.

a) Necessary Conditions.—To obtain these conditions $f(x)$ we assume that $u(x)$ is a continuous function of x satisfying (51). Multiply both sides of (51) by $\bar{\varphi}_\alpha(x)$, where

$$\bar{\varphi}_\alpha(x) = \lambda_o \int_a^b K(t, x)\bar{\varphi}_\alpha(t)dt,$$

and integrate with respect to x from a to b. We obtain

$$(52) \quad \int_a^b f(x)\bar{\varphi}_\alpha(x)dx = \int_a u(x)\bar{\varphi}_\alpha(x)dx - \lambda_o \int_a^b \bar{\varphi}_\alpha(x)\left\{\int_a^b K(x, t)u(t)dt\right\}dx.$$

In the last integral on the right, $\bar{\varphi}_\alpha(x)$ is constant with respect to t and so can be placed under the second sign of integration. We may change the order of integration and then take $u(t)$ from under the sign of integration with respect to x. Thus the last term becomes

$$\int_a^b u(t)\left\{\lambda_o \int_a^b \bar{\varphi}_\alpha(x)K(x, t)dx\right\}dt = \int_a^b u(t)\bar{\varphi}_\alpha(t)dt.$$

Thus we see that the first and last terms on the right cancel and

$$(53) \quad \int_a^b f(x)\bar{\varphi}_\alpha(x)dx = 0. \qquad (\alpha = 1, \ldots, q)$$

Hence, in order that there may exist a continuous solution $u(x)$ of (51), $f(x)$ must satisfy the q conditions (53).

b) Sufficiency Proof.—Let us now show conversely that, if $f(x)$ satisfies the q conditions (53), then (51) does have a solution. By our hypothesis the q equations (53) are satisfied. Then

$$\sum_{\alpha=1}^{q} \lambda_o K(x, y_\alpha') \int_a^b f(t)\bar{\varphi}_\alpha(t)dt = 0.$$

Now, $\lambda_o K(x, y_\alpha')$ is independent of t and so may be placed under the sign of integration. It is permissible to change the order of performing the summation and the integration. We thus obtain

$$\int_a^b \left\{ \sum_{\alpha=1}^{q} \lambda_o K(x, y_\alpha')\bar{\varphi}_\alpha(t)f(t) \right\} dt = 0,$$

which, on account of (47), becomes

$$(54)\quad 0 = \int_a^b \lambda_o K(x, t)f(t)dt - \int_a^b H(x, t)f(t)dt$$
$$+\lambda_o \int_a^b f(t) \left\{ \int_a^b H(s, t)K(x, s)ds \right\} dt.$$

The last term may be written

$$\lambda_o \int_a^b \int_a^b f(t)H(s, t)K(x, s)ds\, dt.$$

Make now a change in notation. In place of t and s write s and t. We obtain

$$\lambda_o \int_a^b \int_a^b f(s)H(t, s)K(x, t)dt\, ds.$$

In this definite double integral it is permissible to change the order of integration. We then obtain

$$\lambda_o \int_a^b K(x, t) \left\{ \int^b H(t, s)f(s)ds \right\} dt.$$

After making these reductions in (54), combine the first and last terms and obtain

$$(55)\quad 0 = \lambda_o \int_a^b K(x, t)\left\{f(t) + \int_a^b H(t, s)f(s)ds\right\}dt - \int_a^b H(x, t)f(t)dt$$

Now put

$$u_o(t) = f(t) + \int_a^b H(t, s)f(s)ds,$$

then

$$\int_a^b H(x, t)f(t)dt = u_o(x) - f(x).$$

Making use of these last two equations, (55) becomes

$$u_o(x) = f(x) + \lambda_o \int_a^b K(x, t)u_o(t)dt.$$

Thus we have proved that if (53) are satisfied then (51) has at least one solution, $u_o(x)$ given by

$$(56)\qquad u_o(x) = f(x) + \int_a^b H(x, t)f(t)dt.$$

c) Determination of All Solutions.—Let us suppose that (51) has another continuous solution $u(x)$. Then $u(x) - u_o(x)$ is a solution of the homogeneous equation

$$(57)\qquad v(x) = \lambda_o \int_a^b K(x, t)v(t)dt,$$

for, if we subtract the members of (56) from the corresponding members of (51), we obtain

$$(58)\quad u(x) - u_o(x) = \lambda_o \int_a^b K(x, t)\Big[u(t) - u_o(t)\Big]dt.$$

By Theorem VIII the most general solution of (57) is of the form

$$C_1\varphi_1(x) + C_2\varphi_2(x) + \cdots + C_q\varphi_q(x),$$

where $C_1, C_2, \ldots, C_q$ are arbitrary constants.

Hence the most general solution of (58) is

$$u(x) - u_o(x) = C_1\varphi_1(x) + C_2\varphi_2(x) + \cdots + C_q\varphi_q(x).$$

Therefore,

$$u(x) = f(x) + \int_a^b H(x, t)f(t)dt + C_1\varphi_1(x) + \ldots + C_q\varphi_q(x)$$

is the complete solution of (51). We have thus proved *Fredholm's third fundamental theorem:*

Theorem XI.—*If λ_o is a characteristic constant of $K(x, t)$ of index q then*

$$(51) \qquad u(x) = f(x) + \lambda_o \int_a^b K(x, t)u(t)dt$$

has, in general, no continuous solution. In order that a continuous solution exist it is necessary that

$$\int_a^b f(x)\bar{\varphi}_\alpha(x)dx = 0, \quad \alpha = 1, \ldots, q$$

where the $\bar{\varphi}_\alpha(x)$ are a complete set of fundamental functions for the associated homogeneous equation

$$(42) \qquad v(x) = \lambda_o \int_a^b K(t, x)v(t)dt.$$

If these conditions are satisfied, then there are a q-fold infinitude of solutions of (51) *given by*

$$u(x) = f(x) + \int_a^b H(x, t)f(t)dt + C_1\varphi_1(x) + \ldots + C_q\varphi_q(x)$$

where $C_1, \ldots, C_q$ are arbitrary constants and where the $\varphi_\alpha(x)$ are a complete set of fundamental functions for

$$(26) \qquad u(x) = \lambda_o \int_a^b K(x, t)u(t)dt \quad \text{(Theorem VIII)}$$

and $H(x, t)$ is given by (40).

The following table exhibits the results of the solution of Fredholm's equation together with the analogy between the finite system and the integral equation.

$$u(x) = f(x) + \lambda \int_a^b K(x, t)u(t)dt$$

Case I: $D(\lambda) \neq 0$		Case II: $D(\lambda) = 0$, index q	
Non-homogeneous	Homogeneous	Non-homogeneous	Homogeneous
Unique solution $u(x) = f(x) + \int_a^b \frac{D(x, t; \lambda)}{D(\lambda)} f(t)dt$	Unique solution $u \equiv 0$	In general, no solution. Solutions exist only if f satisfies $\int_a^b f(x)\bar{\varphi}_\alpha(x) = 0$. Then ∞^q solutions	∞^q solutions: $\sum_{\alpha=1}^{q} C_\alpha \varphi_\alpha(x)$

$$u_i - \lambda h \sum_{j=1}^{n} K_{ij}u_j = f_j,\ i = 1, \ . \ . \ . \ , n$$

Case I: $\Delta \neq 0$		Case II: $\Delta = 0$, index q	
Non-homogeneous	Homogeneous	Non-homogeneous	Homogeneous
Unique solution $u_k = \frac{\sum_{i=1}^{n} f_i \Delta_{ik}}{\Delta}$	Unique solution $u_k = 0$	In general, no solution. ∞^q solutions if the f_i satisfy certain q relations of the form $C_{\alpha 1} f_1 + \ . \ . \ . + C_{\alpha n} f_n = 0$, $\alpha = 1 \ . \ . \ . \ , q$	∞^q solutions if Δ is of rank r where $q = n - r$

25. Kernels of the Form $\sum a_i(x) b_i(y)$.—We give a brief discussion of the integral equation with a kernel of the form

$$K(x, y) = a_1(x)b_1(y) + \dots + a_n(x)b_n(y).$$

Fredholm's integral equation

$$u(x) = f(x) + \lambda \int_a^b K(x, t)u(t)dt$$

can now be written in the form

$$(59)\quad u(x) = f(x) + \lambda\left[a_1(x)\int_a^b b_1(t)u(t)dt + \dots + a_n(x)\int_a^b b_n(t)u(t)dt\right].$$

If, now, we put

$$(60)\quad \int_a^b b_i(t)u(t)dt = K_i \qquad (i = 1, \dots, n),$$

we see that $u(x)$ is of the form

$$(61)\quad u(x) = f(x) + \lambda\left[a_1(x)K_1 + \dots + a_n(x)K_n\right].$$

In order to determine the constants K_i, let us substitute in (60) the value of u given by (61). We obtain the n equations

$$(62)\quad K_i - \lambda\left[\int_a^b a_1(t)b_i(t)K_1dt + \dots + \int_a^b a_n(t)b_i(t)K_ndt\right] = \int_a^b b_i(t)f(t)dt \quad (i = 1, \dots, n).$$

Introducing the notation

$$\int_a^b a_k(t)b_i(t)dt = C_{ki}$$

the system (62) can be written in the form

$$(63)\quad K_i - \lambda\left[C_{1i}K_1 + \dots + C_{ni}K_n\right] = \int_a^b b_i(t)f(t)dt.$$

Equations (63) are a linear algebraic system of n non-homogeneous equations in the n unknowns $K_1, \ldots, K_n$, with the determinant

$$D(\lambda) = \begin{vmatrix} 1 - \lambda C_{11} & -\lambda C_{12} & \cdots & -\lambda C_{1n} \\ -\lambda C_{21} & 1 - \lambda C_{22} & \cdots & -\lambda C_{2n} \\ \cdots & \cdots & \cdots & \cdots \\ \cdots & \cdots & \cdots & \cdots \\ -\lambda C_{n1} & -\lambda C_{n2} & \cdots & 1 - \lambda C_{nn} \end{vmatrix}$$

From the theory of linear systems, we have at once the result:

a) If $D(\lambda) \neq 0$, the system (62) *is satisfied by one and only one set of values of $K_1, \ldots, K_n$ and these values are given by Cramer's formulas. Therefore, Fredholm's equation* (59) *has one and only one solution, which is given by* (61).

b) If $D(\lambda) = 0$ for $\lambda = \lambda_o$ (and this happens for n values of λ, real or complex), and one of the qth minors of $D(\lambda)$ is the first minor which does not vanish for $\lambda = \lambda_o$ (this qth minor is a determinant of order $n - q$), then the general solution of the homogeneous system (62) ($f(x) \equiv 0$) will be of the form[1]

$$K_i = \alpha_1 m_{1i} + \alpha_2 m_{2i} + \cdots + \alpha_q m_{qi} \quad (i = 1, \ldots, n),$$

where $\alpha_1, \alpha_2, \ldots, \alpha_q$ are arbitrary constants.

If we put the values of K_i so obtained in (61), we obtain

$$u(x) = \lambda_o \Big[\alpha_1 u_1(x) + \alpha_2 u_2(x) + \cdots + \alpha_q u_q(x) \Big]$$

where the functions

$$u_r(x) = m_{r1}\, a_1(x) + m_{r2} a_2(x) + \cdots + m_{rn} a_n(x) \qquad (r = 1, \ldots, q)$$

are linearly independent.

Thus we see that, under the circumstances specified, the homogeneous integral equation for $\lambda = \lambda_o$ has q linearly independent solutions.

[1] BÖCHER, "Introduction to Higher Algebra," §18.

The *associated equation*

$$(64)\quad \bar{u}(x) = f(x) + \lambda\left[b_1(x)\int_a^b a_1(t)\bar{u}(t)dt + \cdots + b_n(x)\int_a^b a_n(t)\bar{u}(t)dt\right]$$

is obtained from (59) by interchanging the functions $a_i(x)$ and $b_i(x)$. The general term of the characteristic determinant of (59) being

$$\int_a^b a_k(t)b_i(t)dt = C_{ki},$$

the general term for the associated equation will be

$$\int_a^b b_k(t)a_i(t)dt = C_{ik}.$$

The characteristic determinants of these two equations are identical, since one can obtain one from the other by interchanging rows and columns. Therefore, the equation (59) and the associated equation (64) have exactly the same characteristic numbers and with the same index.

From the general theory we know that if λ_o is a root of $D(\lambda) = 0$ of index q, then, in order that the non-homogeneous equation (59) may have a solution, we must have

$$\int_a^b f(x)\bar{u}_i(x)\,dx = 0 \quad (i = 1, \ldots, q).$$

EXERCISES

For the equation $u(x) = f(x) + \lambda\int_a^b K(x, t)u(t)dt$ compute $D(\lambda)$ and $D(x, y; \lambda)$ for the following kernels for the specified limits a and b:

	Ans.
1. $K(x, t) = 1,\ a = 0,\ b = 1.$	$D(\lambda) = 1 - \lambda.$
2. $K(x, t) = -1,\ a = 0,\ b = 1.$	$D(\lambda) = 1 + \lambda.$
3. $K(x, t) = \sin x,\ a = 0,\ b = \pi.$	$D(\lambda) = 1 - 2\lambda.$
4. $K(x, t) = xt,\ a = 0,\ b = 10.$	$D(\lambda) = 1 - \frac{1{,}000}{3}\lambda.$

5. $K(x, t) = t,\ a = 0,\ b = 10.$ $\qquad D(\lambda) = 1 - 50\lambda.$

6. $K(x, t) = x,\ a = 4,\ b = 10.$ $\qquad D(\lambda) = 1 - 42\lambda.$

7. $K(x, t) = g(x),\ a = a,\ b = b.$ $\qquad D(\lambda) = 1 - \lambda \int_a^b g(t)dt.$

8. $K(x, t) = g(t),\ a = a,\ b = b.$ $\qquad D(\lambda) = 1 - \lambda \int_a^b g(t)dt.$

9. $K(x, t) = 2e^x \cdot e^t,\ a = 0,\ b = 1.$ $\qquad D(\lambda) = 1 - (e^2 - 1)\lambda.$

10. $K(x, t) = x - t,\ a = 0,\ b = 1.$

Solve the following integral equations:

11. $u(x) = \sec^2 x + \lambda \int_0^1 u(t)dt.$

12. $u(x) = \sec x \cdot \tan x - \lambda \int_0^1 u(t)dt.$

13. $u(x) = \cos x + \lambda \int_0^\pi \sin x \cdot u(t)dt.$

14. $u(x) = e^x + \lambda \int_0^{10} xt \cdot u(t)dt.$

15. $u(x) = x^2 + \lambda \int_0^{10} t \cdot u(t)dt.$

16. $u(x) = \sin x + \lambda \int_4^{10} x \cdot u(t)dt.$

17. $u(x) = e^x + \lambda \int_0^1 2e^x e^t u(t)dt.$

Solve the following homogeneous integral equations:

18. $u(x) = \int_0^1 u(t)dt.$

19. $u(x) = \int_0^1 (-1)u(t)dt.$

20. $u(x) = \frac{1}{2}\int_0^\pi \sin x . u(t)dt.$

21. $u(x) = \frac{3}{1{,}000}\int_0^{10} xt . u(t)dt.$

22. $u(x) = \frac{1}{50}\int_0^{10} t . u(t)dt.$

23. $u(x) = \frac{1}{42}\int_4^{10} x.u(t)dt.$

24. $u(x) = \frac{1}{e^2 - 1}\int_0^1 2e^x e^t u(t)dt.$

Solve the following equation by the method of §25:

25. $u(x) = x^2 + \lambda \int_0^1 (1 + xt)u(t)dt.$

26. $u(x) = x + \lambda \int_0^\pi (1 + \sin x \sin t)u(t)dt.$

27. $u(x) = x + \lambda \int_0^1 (1 + x + t)u(t)dt.$

28. $u(x) = x + \lambda \int_0^1 (x - t)u(t)dt.$

29. $u(x) = x + \lambda \int_0^1 (x - t)^2 u(t)dt.$

30. Solve Exercises 11–17 inclusive by this method.

CHAPTER IV

APPLICATIONS OF THE FREDHOLM THEORY

I. Free Vibrations of an Elastic String

26. The Differential Equations of the Problem.—We consider an elastic string stretched between the two fixed points A and B. We pull it out of its position of equilibrium AQB into some other plane initial position, as ACB, and then release it. The string will describe transverse vibrations. Suppose that at time t the string occupies the position APB. Let $x = AQ$, $y = QP$ be the abscissa and ordinate of any one of its points P. Then y is a function of x and t. We suppose the cross-section of the string to be constant and infinitesimal compared with the length.

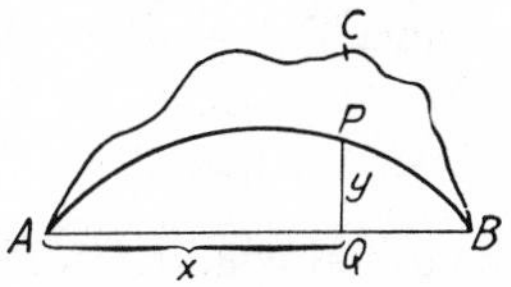

Fig. 5.

The string is of homogeneous density. The effect of gravity is to be neglected. Further, we take for simplicity $AB = 1$. It is then proved in the theory of elasticity that the motion of the string is given by the partial differential equation

$$\text{(1)} \qquad \frac{\partial^2 y}{\partial t^2} = c^2 \frac{\partial^2 y}{\partial x^2}, \quad (c = \text{constant})$$

with the initial conditions

$$\text{(2)} \qquad y(0, t) = 0, \qquad y(1, t) = 0,$$
$$\text{(3)} \qquad y(x, 0) = g(x), \qquad y_t(x, 0) = 0,$$

where $y = g(x)$ is the equation of the initial position ACB. Equations (2) are the analytic statement of the fact that the end points A and B remain fixed, while equations (3) state that for $t = 0$ the string is in the initial position ACB and each particle starts its motion with an initial zero velocity.[1]

27. Reduction to a One-dimensional Boundary Problem. Let us try to find a solution of (1) in the form

$$y = u(x)\varphi(t).$$

Substitution of this value of y in (1) gives

$$u(x)\frac{d^2\varphi}{dt^2} = c^2\varphi(t)\frac{d^2u}{dx^2},$$

which can be put in the form

$$\frac{\dfrac{d^2\varphi(t)}{dt^2}}{\varphi(t)} = c^2\frac{\dfrac{d^2u(x)}{dx^2}}{u(x)}.$$

The right-hand side is independent of t, and the left-hand side is independent of x. Then either member is a constant, which we designate as $-\lambda c^2$. This gives us the two ordinary differential equations to solve:

$$\frac{d^2u}{dx^2} + \lambda u = 0, \qquad \frac{d^2\varphi}{dt^2} + \lambda c^2\varphi = 0.$$

The initial conditions on u and φ are obtained from (2) and (3), and for u are as follows:

$$\begin{matrix} u(0)\varphi(t) = 0 \\ u(1)\varphi(t) = 0 \end{matrix} \quad \text{whence} \quad \begin{matrix} u(0) = 0 \\ u(1) = 0 \end{matrix} \quad \text{since } \varphi(t) \not\equiv 0.$$

We are thus led to the following *one-dimensional boundary problem*: to determine a function $u(x)$ which will satisfy the differential equation

$$\frac{d^2u}{dx^2} + \lambda u = 0 \tag{4}$$

and the boundary conditions

$$u(0) = 0, \qquad\qquad u(1) = 0.$$

28. Solution of the Boundary Problem.—We see at once that $u \equiv 0$ is a solution. This gives for the original prob-

[1] WEBER-RIEMANN, "Lehrbuch der Partielle Differentialgleichungen," §83.

lem of the vibrating string the trivial solution $y \equiv 0$, which means that the string remains at rest. Hereafter, when we refer to a solution of our boundary problem we shall mean a solution not identically zero. To obtain the solution we have three cases to consider.

Case I.—$\lambda > 0$. From the elementary theory of differential equations we know that the most general solution of (4) is

$$u(x) = A \cos \sqrt{\lambda}x + B \sin \sqrt{\lambda}\cdot x.$$

The conditions $u(0) = 0$, $u(1) = 0$ give us $A = 0$ and either (1) $B = 0$, or (2) $\sin \sqrt{\lambda} = 0$.

(1) If $B = 0$, we obtain the trivial solution $u \equiv 0$.

(2) If $\sin \sqrt{\lambda} = 0$, then $\lambda = n^2\pi^2$ (n, an integer) and the solution is

$$u(x) = B \sin n\pi x$$

Case II.—$\lambda = 0$. The general solution of (4) is now

$$u = Ax + B.$$

But $u(0) = B = 0$, and $u(1) = A = 0$, so that we have again the trivial solution $u \equiv 0$.

Case III.—$\lambda < 0$. The general solution of (4) is now

$$u = Ae^{\sqrt{-\lambda}x} + Be^{-\sqrt{-\lambda}x}.$$

Applying the initial conditions, we obtain

$$u(0) = A + B = 0, \text{ whence } B = -A$$

$$u(1) = A(e^{\sqrt{-\lambda}} - e^{-\sqrt{-\lambda}}) = 0, \text{ whence } A = 0\ (\lambda \neq 0).$$

Therefore, $A = B = 0$, and we have again the trivial solution $u \equiv 0$.

Thus we arrive at the

Theorem I.—*If $\lambda = n^2\pi^2$(n, an integer), then the boundary problem*

$$\frac{d^2u}{dx^2} + \lambda u = 0,\ u(0) = 0,\ u(1) = 0 \tag{4}$$

has an infinitude of solutions:

$$u = B \sin n\pi x.$$

If $\lambda \neq n^2\pi^2$, *then the only solution is the trivial one* $u \equiv 0$.

29. Construction of Green's Function.—We now propose to show that every solution of our boundary problem (4) satisfies at the same time a linear integral equation. We observe first that with the given boundary conditions the method of §4 cannot be used to determine an equivalent integral equation. In the present instance, in order to determine an equivalent integral equation, we first construct the *Green's function* belonging to the boundary problem.

The given boundary problem for $\lambda = 0$ has only the trivial solution $u \equiv 0$. This is true, however, only under the assumptions tacitly made throughout, namely, that u, together with its first and second derivatives, is continuous in the interval [01]. Let us use the notation $u^{c''}$ to denote this assumption.[1]

Drop now the assumption $u^{c''}$ and allow the derivative u' to be discontinuous at an arbitrarily prescribed point ξ between 0 and 1, while u itself remains continuous. Accordingly, we propose to determine a function u satisfying the following conditions:

A) u^c in [01].

B) $u^{c''}$ and $\dfrac{d^2u}{dx^2} = 0$ in $[0\xi]$.

$u^{c''}$ and $\dfrac{d^2u}{dx^2} = 0$ in $[\xi 1]$.

C) $u(0) = 0,\ u(1) = 0$.

The solution then must be of the form

$$u = \begin{cases} \alpha_0 x + \beta_0 \text{ in } [0\xi] \\ \alpha_1 x + \beta_1 \text{ in } [\xi 1]. \end{cases}$$

[1] Similarly, let us use the notation C to denote the class of all continuous functions and the notation $C^{(n)}$ to denote the class of all functions having continuous derivatives up to the order n inclusive.

From C) we find

$$u(0) = \beta_0 = 0; \; u(1) = \alpha_1 + \beta_1 = 0.$$

Thus the solution reduces to

$$u = \begin{cases} \alpha_0 x & [0\xi] \\ \beta_1(1 - x) & [\xi 1]. \end{cases}$$

The condition A) must be satisfied, whence

$$\alpha_0 \xi = \beta_1(1 - \xi).$$

Therefore $\qquad \alpha_0 = \rho(1 - \xi)$ and $\beta_1 = \rho\xi.$

The solution now takes the form

$$u = \begin{cases} \rho(1 - \xi)x & [0\xi] \\ \rho\xi(1 - x) & [\xi 1]. \end{cases}$$

Geometrically, this solution is represented by a broken line as in the adjoining figure.

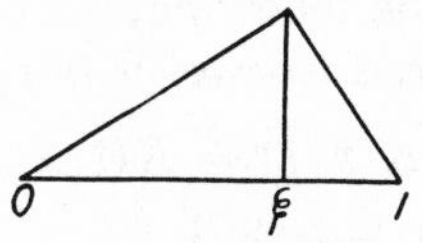

FIG. 6.

For $x = \xi$, the derivative has a discontinuity measured by

$$u'(\xi - 0) - u'(\xi + 0) = \rho(1 - \xi) + \rho\xi = \rho.$$

We now impose the further condition that this discontinuity shall be $+1$. Then $\rho = 1$. The function u so obtained is called *Green's function* for the boundary problem (4). We use the notation $K(x, \xi)$ to represent this function. We have thus the following:

Theorem II.—*There exists one and only one function* $K(x, \xi)$ *which satisfies the conditions*

$A)$ K^c in $[01]$.

$B)$ $K^{c''}$ and $\dfrac{d^2K}{dx^2} = 0$ in $[0\xi]$.

$K^{c''}$ and $\dfrac{d^2K}{dx^2} = 0$ in $[\xi 1]$.

$C) K(0) = 0,\ K(1) = 0.$

$$D)\ \Big[K'\Big]_{x = \xi + 0}^{x = \xi - 0} = +1, \quad 0 < \xi < 1.$$

This function is given by the formula

$$K(x, \xi) = \begin{cases} (1 - \xi)x \text{ for } 0 \leqq x \leqq \xi \\ \xi(1 - x) \text{ for } \xi \leqq x \leqq 1. \end{cases} \tag{5}$$

Let us use the notation

$$K_0(x, \xi) = (1 - \xi)x$$
$$K_1(x, \xi) = \xi(1 - x).$$

The properties $A)$ and $D)$ of $K(x, \xi)$ may now be written

$$K_0(\xi, \xi) = K_1(\xi, \xi)$$
$$K_0'(\xi, \xi) - K_1'(\xi, \xi) = 1.$$

We can now prove the following:

Theorem III.—*Green's function is symmetric in x and ξ:*

$$K(x, \xi) = K(\xi, x).$$

Proof.—Let $0 \leqq z_1 \leqq z_2 \leqq 1$.

Then $\quad K(z_1, z_2) = K_0(z_1, z_2) = (1 - z_2)z_1$

while $\quad K(z_2, z_1) = K_1(z_1, z_2) = z_1(1 - z_2)$

whence $\quad K(z_1, z_2) = K(z_2, z_1)$.

30. Equivalence between the Boundary Problem and a Linear Integral Equation.—Take the equations

$$\frac{d^2u}{dx^2} = -\lambda u, \quad \frac{d^2K}{dx^2} = 0.$$

Multiply the first by $-K$ and the second by u and add. We obtain

$$uK'' - Ku'' = \lambda uK, \text{ or}$$
$$\frac{d}{dx}(uK' - Ku') = \lambda uK.$$

This equality holds in each of the two subintervals [0ξ] and [ξ1]. Integration over each of them gives

$$\left[uK' - Ku'\right]_0^{\xi-0} = \lambda \int_0^{\xi-0} uKdx$$

$$\left[uK' - Ku'\right]_{\xi+0}^{1} = \lambda \int_{\xi+0}^{1} uKdx.$$

Both u and K vanish at 0 and 1. u, u', and K are continuous functions of x over the whole interval [01]. Whence, adding the last two equations, we obtain

$$u(\xi)\left[K'(\xi - 0) - K'(\xi + 0)\right] = \lambda \int_0^1 K(x, \xi)u(x)dx.$$

But $K'(\xi - 0) - K'(\xi + 0) = 1$, and so

$$u(\xi) = \lambda \int_0^1 K(x, \xi)u(x)dx.$$

Now interchange x and ξ and remember that $K(x, \xi) = K(\xi, x)$. Then

$$u(x) = \lambda \int_0^1 K(x, \xi)u(\xi)d\xi.$$

This is a homogeneous linear integral equation of the second kind for the determination of $u(x)$. Every solution of the original boundary problem (4) satisfies this integral equation. Hence, we have the following:

Theorem IV.—*If $u(x)$ has continuous first and second derivatives, and satisfies the boundary problem*

$$\text{(4)} \qquad \frac{d^2u}{dx^2} + \lambda u = 0, \; u(0) = 0, \; u(1) = 0,$$

then $u(x)$ is continuous and satisfies the homogeneous linear integral equation

$$u(x) = \lambda \int_0^1 K(x, \xi)u(\xi)d\xi,$$

where $K(x, \xi)$ is given by the formula (5).

Let us now prove the following converse

Theorem V.—*If $u(x)$ is continuous and satisfies the equation*

$$u(x) = \lambda \int_0^1 K(x, \xi)u(\xi)d\xi, \tag{6}$$

where $K(x, \xi)$ is given by the formula (5), *then u has continuous first and second derivatives and satisfies the boundary problem*

$$\frac{d^2u}{dx^2} + \lambda u = 0, \; u(0) = 0, \; u(1) = 0.$$

Proof.—x and ξ range from 0 to 1 and $K'(x, \xi)$ is discontinuous at $x = \xi$. Let us then write (6) in the form

$$u(x) = \lambda \int_0^x K_1(x, \xi)u(\xi)d\xi + \lambda \int_x^1 K_0(x, \xi)u(\xi)d\xi.$$

Now we may apply the general rule for the differentiation of a definite integral with respect to a parameter.[1] Then

$$\frac{du}{dx} = \lambda \int_0^x K_1'(x, \xi)u(\xi)d\xi + \lambda \int_x^1 K_0'(x, \xi)u(\xi)d\xi.$$

since
$$K_0(x, x) = K_1(x, x).$$

Moreover, since $K_1'(x, \xi)$ and $K_0'(x, \xi)$ are continuous in their respective intervals, we see that $\dfrac{du}{dx}$ is continuous. A second differentiation gives

$$\frac{d^2u}{dx^2} = \lambda \int_0^x K_1''(x, \xi)u(\xi)d\xi + \lambda \int_x^1 K_0''(x, \xi)u(\xi)d\xi$$
$$+ \lambda K_1'(x, x)u(x) - \lambda K_0'(x, x)u(x) = -\lambda u(x),$$

since by our hypothesis on K we have $K_1'' = K_0'' = 0$ and $K_1'(x, x) - K_0'(x, x) = -1$. Moreover, since $u(x)$ is continuous, we have u'' continuous. We have further

$$u(0) = \lambda \int_0^1 K(0, \xi)u(\xi)d\xi = 0,$$

[1] See Goursat-Hedrick, "Mathematical Analysis," vol. 1, §97.

since $K(0, \xi) = 0$, and

$$u(1) = \lambda \int_0^1 K(1, \xi)u(\xi)d\xi = 0,$$

since $K(1, \xi) = 0$.

We know the solution of the given boundary problem (4). This knowledge combined with Theorems IV and V gives the following:

Theorem VI.—*Only when* $\lambda = n^2\pi^2(n,$ *an integer*) *does the integral equation*

$$u(x) = \lambda \int_0^1 K(x, \xi)u(\xi)d\xi$$

have a solution not identically zero:

$$u(x) = B \sin n\pi x.$$

If we compare these results with the results obtained in the preceding chapter for the general homogeneous integral equation, we see that the characteristic constants for this particular problem are $\lambda = n^2\pi^2 \equiv \lambda_n$ and that they are of index $q = 1$. The fundamental function belonging to λ_n is

$$\varphi(x) = \sin n\pi x.$$

The kernel is a symmetric one, so that

$$K(x, \xi) = K(\xi, x),$$

and therefore the associated equation is identical with the original one and hence has the same solutions. The associated fundamental functions are, therefore,

$$\bar{\varphi}(x) = \sin n\pi x.$$

II. Constrained Vibrations of an Elastic String

31. The Differential Equations of the Problem.—Let us suppose that an exterior force $\mu H(x, t)$ acts on each particle of mass μ in the y direction. Then it is known from the mathematical theory[1] of vibrating strings that the equations of motion of the string are

$$\frac{\partial^2 y}{\partial t^2} = c^2 \frac{\partial^2 y}{\partial x^2} + H(x, t), \tag{7}$$

$$y(0, t) = 0, \qquad y(1, t) = 0, \tag{8}$$

$$y(x, 0) = g(x), \qquad y_t(x, 0) = 0. \tag{9}$$

Let us suppose now that $H(x, t)$ is harmonic, that is,

$$H(x, t) = C^2 r(x) \cos (\beta t + \gamma) \quad (c \neq 0).$$

Let us find, if possible, a solution of the form

$$y(x, t) = u(x) \cos (\beta t + \gamma).$$

Substitute this value of y in (7) and put $C^2\lambda = \beta^2$. We find

$$\frac{d^2u}{dx^2} + \lambda u + r(x) = 0, \tag{10}$$

while from (8) we derive the boundary conditions

$$u(0) = 0, \quad u(1) = 0. \tag{11}$$

32. Equivalence Between the Boundary Problem and a Linear Integral Equation.—Construct as before the Green's function $K(x, \xi)$. Then

$$\frac{d^2K}{dx^2} = 0, \qquad \frac{d^2u}{dx^2} = -\lambda u - r.$$

Multiply the first of these by u, the second by $-K$ and add. We obtain

$$uK'' - Ku'' = \lambda uK + rK,$$

which may be written

$$\frac{d}{dx}(uK' - Ku') = \lambda uK + rK.$$

Proceed as before with the integration from 0 to ξ and from ξ to 1. We find

$$u(\xi) = \lambda \int_0^1 K(x, \xi) u(x) dx + \int_0^1 K(x, \xi) r(x) dx.$$

[1] WEBER, *Loc. cit.*, §83.

Interchange x and ξ. Then, on account of the symmetry of K, we have

$$u(x) = \lambda \int_0^1 K(x, \xi)u(\xi)d\xi + \int_0^1 K(x, \xi)r(\xi)d\xi.$$

If, now, we put

$$\int_0^1 K(x, \xi)r(\xi)d\xi = f(x) \tag{12}$$

we have

$$u(x) = f(x) + \lambda \int_0^1 K(x, \xi)u(\xi)d\xi, \tag{13}$$

which is a non-homogeneous linear integral equation of the second kind. It is satisfied by every solution of the boundary problem given by (10) and (11).

If we will proceed exactly as in the case of the homogeneous equation, we can now show conversely that if u is continuous and satisfies the equation (13), where $f(x)$ is given by (12), then u has continuous first and second derivatives and satisfies the differential equation (10) and the boundary conditions (11).

33. Remarks on Solution of the Boundary Problem.—Equations (6) and (13) have the same kernel, namely, the Green's function which we have constructed. Knowing that the characteristic constants for (6), and hence for (13), are $\lambda_n = n^2\pi^2$, we obtain from the general theory the following results for (13):

Case I.—If $\lambda \neq n^2\pi^2$, (13) has a unique solution.

Case II.—If $\lambda = n^2\pi^2$, there is, in general, no solution. A solution exists only when the condition

$$\int_0^1 f(x) \sin n\pi x dx = 0 \tag{14}$$

is satisfied. This condition is what

$$\int_a^b f(x)\bar{\varphi}_\alpha(x)dx = 0$$

becomes for this special problem. If (14) is satisfied, then (13) has ∞^1 solutions.

The boundary problem

$$\frac{d^2u}{dx^2} + \lambda u + r(x) = 0,\ u(0) = 0,\ u(1) = 0$$

was shown to be equivalent to the non-homogeneous integral equation (13) if

$$f(x) = \int_0^1 K(x, \xi) r(\xi) d\xi. \tag{15}$$

This enables us, from Cases I and II for the integral equation, to state that, when $\lambda \neq n^2\pi^2$, the boundary problem has a unique solution, and when $\lambda = n^2\pi^2$, there is, in general, no solution, but that when (14) is satisfied, there are ∞^1 solutions. If, now, in (14) we substitute for $f(x)$ its value as given by (15), this condition becomes

$$\int_0^1 \int_0^1 K(x, \xi) r(\xi) \sin n\pi x d\xi dx = 0.$$

Interchange the order of integration and remember that $\sin n\pi x$ is a solution of the homogeneous equation for $\lambda = \lambda_n$:

$$\sin n\pi x = \lambda_n \int_0^1 K(x, \xi) \sin n\pi \xi d\xi$$

or

$$\sin n\pi \xi = \lambda_n \int_0^1 K(x, \xi) \sin n\pi x dx$$

[since $K(x, \xi) = K(\xi, x)$], then the double integral becomes

$$\frac{1}{\lambda_n} \int^1 r(\xi) \sin n\pi \xi d\xi = 0.$$

That is, in order that the boundary problem (10) may have a solution, it is necessary and sufficient that $r(x)$ satisfies the equation

$$\int_0^1 r(x) \sin n\pi x dx = 0.$$

III. Auxiliary Theorems on Harmonic Functions

34. Harmonic Functions.—For the solution later of the two boundary problems of the potential theory known as *Dirichlet's problem* and *Neumann's problem* we shall need certain auxiliary theorems on *harmonic functions,* that is, functions of class C'' satisfying $\Delta u = 0$:

$$\Delta u \equiv \frac{\partial^2 u}{\partial x^2} + \frac{\partial^2 u}{\partial y^2} = 0.$$

35. Definitions about Curves.—A curve C

$$C: \qquad x = \varphi(t), \quad y = \psi(t), \quad t_o \leqq t \leqq t_1$$

is said to be *continuous* if φ and ψ are continuous on $[t_o, t_1]$. We write this

$$C^c \cdot\!\sim\!\cdot \varphi^c \cdot \psi^c.$$

The symbol $\cdot\!\sim\!\cdot$ is read *is equivalent to* or *implies and is implied by.* The symbol $\cdot$ is read *and.* Further, we say that the curve C is of class c' if φ and ψ have continuous first derivatives and these first derivatives φ' and ψ' do not vanish simultaneously on $[t_o, t_1]$. We write this

$$C^{c'} \cdot\!\sim\!\cdot \varphi^{c'} \cdot \psi^{c'} \cdot (\varphi', \psi') \neq (0, 0).$$

Such a curve is sometimes called a *smooth curve.* A curve of class c' has a definite positive tangent at every point. The condition $(\varphi', \psi') \neq (0, 0)$ excludes singular points. Every arc has a definite finite length. Thus we are assured that for smooth curves we can choose as parameter t the length s of arc and write

$$C: \qquad x = \xi(s), \quad y = \eta(s), \quad 0 \leqq s \leqq l$$
$$\xi'^2 + \eta'^2 = 1,$$

when l is the total length of arc.

In like manner, we define a curve of class C'':

$$C^{c''} \cdot \sim \cdot \varphi^{c''} \cdot \psi^{c''} \cdot (\varphi', \psi') \neq (0, 0).$$

Curves of class c'' have at every point a definite curvature,[1] which varies continuously from point to point.

36. Green's Theorem.—Let C be a curve with the following properties:

1) It is closed.
2) It has no multiple points.
3) It is of class c''. Hence we can represent it with the arc s as parameter:

$$x = \xi(s), \quad y = \eta(s), \quad 0 \leqq s \leqq l.$$

4) There exists a positive integer m such that every line parallel to the y or x axis meets the curve in at most m points.

According to Jordan's theorem[2] such a curve divides the plane into an interior and an exterior region. Let us denote by

I, the interior plus the boundary C,
by E, the exterior plus the boundary C.

Then Green's theorem[3] may be stated as follows:

Green's Theorem.—*If $P(x, y)$ and $Q(x, y)$ are of class c' on I, then*

$$(16) \quad \int_{(C)} \left[P(x, y)dx + Q(x, y)dy \right] = \iint_I \left(\frac{\partial Q}{\partial x} - \frac{\partial P}{\partial y} \right) dxdy$$

where the line integral is taken in the positive sense around C.

[1] See GOURSAT-HEDRICK, "Mathematical Analysis," vol. 1, §205, note.

[2] See OSGOOD, "Funktionentheorie," 2nd ed., p. 171.

[3] See GOURSAT-HEDRICK, "Mathematical Analysis," vol. 1, §126, for a proof of Green's theorem.

If we put $P = v\frac{\partial u}{\partial y}$, $Q = -v\frac{\partial u}{\partial x}$, then Green's theorem becomes: If v is of class c' and u of class c'' on I, then

$$\int_{(C)} v\left(\frac{\partial u}{\partial y}\,dx - \frac{\partial u}{\partial x}dy\right) = -\int\int_I v\Delta u dxdy - \int\int_I \left(\frac{\partial u}{\partial x}\frac{\partial v}{\partial x} + \frac{\partial u}{\partial y}\frac{\partial v}{\partial y}\right)dxdy. \tag{17}$$

The left-hand side can be simplified if we introduce the idea of the *directional derivative* of a function $f(x, y)$.

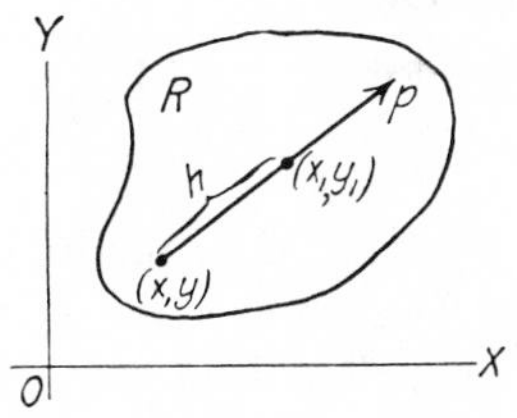

FIG. 7.

Let $f(x, y)$ be defined at every point of a region R of the xy-plane. The directional derivative is defined[1] as follows (see figure):

$$\frac{\partial f}{\partial p} = \lim_{h \to 0} \frac{f(x_1, y_1) - f(x, y)}{h},$$

if such limit exists.

If $f(x, y)$ is of class c', then[2]

$$f(x_1, \ y_1) - f(x, \ y) = f_x(x, \ y)(x_1 - x) + f_y(x, \ y)(y_1 - y) + \alpha(x_1 - x) + \beta(y_1 - y),$$

where α and β approach 0 with $x_1 - x$ and $y_1 - y$.

Divide both members of this last equality by h and then let h approach zero. We obtain

$$\frac{\partial f(x, y)}{\partial p} = \frac{\partial f}{\partial x} \cos (px) + \frac{\partial f}{\partial y} \cos (py). \tag{18}$$

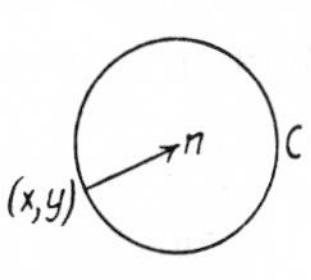

FIG. 8.

Let now n denote the direction of the inner normal to our curve C at a point (x, y) of C. Then, according to (18),

$$\frac{\partial u}{\partial n} = \frac{\partial u}{\partial x} \cos (nx) + \frac{\partial u}{\partial y} \cos (ny).$$

[1] See OSGOOD, "Differential and Integral Calculus," rev. ed., p. 308, 1910.

[2] See OSGOOD, *Loc. cit.*, p. 292.

But

$$\cos(nx) = -\frac{dy}{ds}, \quad \cos(ny) = \frac{dx}{ds},$$

then

$$\frac{\partial u}{\partial n} = -\frac{\partial u}{\partial x}\frac{\partial y}{ds} + \frac{\partial u}{\partial y}\frac{dx}{ds}.$$

Thus, in Green's theorem, we may put

$$\int_{(C)} v\left(\frac{\partial u}{\partial y}\frac{dx}{ds} - \frac{\partial u}{\partial x}\frac{dy}{ds}\right)ds = \int_{(C)} v\frac{\partial u}{\partial n}ds.$$

We have then the following theorem:

Theorem VII.—*If u belongs to the class c'' and v to the class c' on I, then*

$$(19) \quad \int_{(C)} v\frac{\partial u}{\partial n}ds = \iint_I v\Delta u\,dxdy - \iint_I \left(\frac{\partial u}{\partial x}\frac{\partial v}{\partial x} + \frac{\partial u}{\partial y}\frac{\partial v}{\partial y}\right)dxdy.$$

We now apply this theorem to two special cases:

Case I.—$v = 1$, $\Delta u = 0$ on I. Then we have the theorem:

Theorem VIII.—*If u is harmonic on I, then*

$$(20) \qquad \int_{(C)} \frac{\partial u}{\partial n}ds = 0.$$

Case II.—$v = u$, $\Delta u = 0$ on I. Then we have the

Theorem IX.—*If u is harmonic on I, then*

$$-\int_{(C)} u\frac{\partial u}{\partial n}ds = \iint_I \left\{\left(\frac{\partial u}{\partial x}\right)^2 + \left(\frac{\partial u}{\partial y}\right)^2\right\}dxdy.$$

Corollary I.—*If u is harmonic on I and vanishes along the boundary of C, then $u \equiv 0$ on I.*

For, by these hypotheses,

$$= 0 \int\int\left\{\left(\frac{\partial u}{\partial x}\right)^2 + \left(\frac{\partial u}{\partial y}\right)^2\right\} dxdy,$$

hence $\dfrac{\partial u}{\partial x} \equiv 0$, $\dfrac{\partial u}{\partial y} \equiv 0$ and, therefore, $u =$ constant on I. But u is continuous and vanishes on the boundary C of I, hence $u \equiv 0$.

Corollary II.—If u is harmonic on I and $\dfrac{\partial u}{\partial n} = 0$ on the boundary C of I, then $u =$ constant on I.

Notice that in this case we cannot draw the conclusion $u \equiv 0$.

37. The Analogue of Theorem IX for the Exterior Region. Under the assumption that u is harmonic on E, it follows that u is harmonic in a region E_o exterior to C and interior to a circle S around the origin which includes C. From Theorem IX we would then have

$$(21) \quad \int\int_{E_o}\left[\left(\frac{\partial u}{\partial x}\right)^2 + \left(\frac{\partial u}{\partial y}\right)^2\right] dxdy = -\int_{\overleftarrow{C}} u\frac{\partial u}{\partial n_e} ds - \int_{\overleftarrow{S}} u\frac{\partial u}{\partial n} ds,$$

where the normals must be drawn toward the interior of the region E_o. We now let the radius r of the circle S approach ∞ and examine the limits of the three terms in (21). The first of the single integrals remains unchanged. In order to see what happens to the second, make a transformation to polar coordinates. Then $u(x, y)$ becomes $U(r, \theta)$, $ds = rd\theta$, and $\dfrac{\partial u}{\partial n} = -\dfrac{\partial u}{\partial r}$, since the normal is opposite in direction to the radius r. Then the second single integral becomes

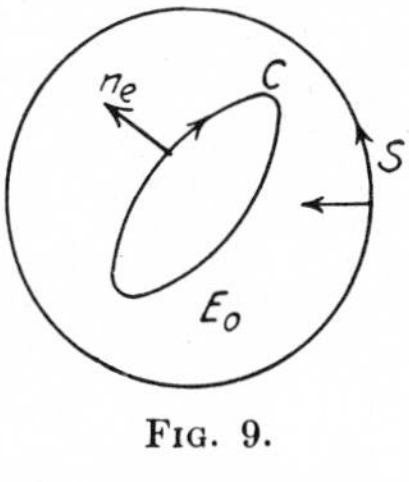

FIG. 9.

$$- \int_0^{2\pi} rU \frac{\partial U}{\partial r} d\theta.$$

Let us now add the assumption that

$$\lim_{r \to \infty} rU \frac{\partial U}{\partial r} = 0, \text{ uniformly with respect to } \theta.$$

Then $-\int_0^{2\pi} rU \frac{\partial U}{\partial r} d\theta$ approaches 0 as r approaches ∞, and, accordingly, also the double integral on the left-hand side of (21) approaches a determinate limit denoted by $\iint\limits_E \left[\left(\frac{\partial u}{\partial x}\right)^2 + \left(\frac{\partial u}{\partial y}\right)^2\right] dxdy.$ Hence we have the theorem:

Theorem X.—*If u is harmonic on E and*

$$\text{(22)} \quad \lim_{r \to \infty} rU \frac{\partial U}{\partial r} = 0, \text{ uniformly with respect to } \theta, \text{ then}$$

$$\text{(23)} \quad \iint\limits_E \left[\left(\frac{\partial u}{\partial x}\right)^2 + \left(\frac{\partial u}{\partial y}\right)^2\right] dxdy = -\int\limits_C u \frac{\partial u}{\partial n_e} ds.$$

As before, we obtain the two corollaries:

Corollary I.—If u is harmonic on E and (22) *still holds, and, moreover, $u = 0$ on C, then $u \equiv 0$ on E.*

Corollary II.—If u is harmonic on E and (22) *still holds, and, moreover, $\frac{\partial u}{\partial n_e} = 0$ on C, then $u =$ constant on E.*

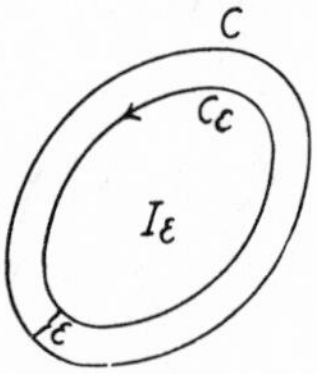

FIG. 10.

38. Generalization of the Preceding.—In the sequel we shall need a modification of the preceding theorems for the case where u is harmonic on the interior but not upon the boundary of C. Let us designate the interior without the boundary C by I'. Construct a closed curve C_ϵ.

$$C_\epsilon: \qquad x = \xi(s, \epsilon), \quad y = \eta(s, \epsilon)$$

of the same character as C interior to C and such that

$$\lim_{\epsilon \to 0} C_\epsilon = C, \text{ uniformly as to } s.$$

Denote by I_ϵ the interior of C_ϵ plus the boundary C_ϵ. Theorem IX applies for this region and we have

$$\iint_{I_\epsilon}\left[\left(\frac{\partial u}{\partial x}\right)^2 + \left(\frac{\partial u}{\partial y}\right)^2\right]dxdy = -\int_{\overset{\frown}{C}_\epsilon} u\frac{\partial u}{\partial n}\,ds.$$

Let us now impose the following three conditions (A):

1) $\lim\limits_{\epsilon\to 0} u = u_i$, uniformly as to s.

2) $\lim\limits_{\epsilon\to 0}\dfrac{\partial u}{\partial n} = \dfrac{\partial u_i}{\partial n}$, uniformly as to s.

3) $\left|\dfrac{\partial u}{\partial x}\right|, \left|\dfrac{\partial u}{\partial y}\right|$ are bounded on I'.

Then we obtain the following theorem:

Theorem XI.—*If u is harmonic on I' and satisfies the three conditions (A), then*

$$\iint_{I}\left[\left(\frac{\partial u}{\partial x}\right)^2 + \left(\frac{\partial u}{\partial y}\right)^2\right]dxdy = -\int_{\overset{\frown}{C}} u_i\frac{\partial u_i}{\partial n}\,ds. \tag{24}$$

The following two corollaries follow as before.

Corollary I.—If we add to the hypotheses of the theorem that $u_i = 0$ along C, then $u \equiv 0$ on I'.

Corollary II.—If we add to the hypotheses of the theorem that $\dfrac{\partial u_i}{\partial n} = 0$ along C, then $u =$ constant on I'.

Make hypotheses similar to (A) for the exterior and call them (B). Then we get for the exterior minus the boundary C, which we designate by E' a corresponding theorem with two corollaries.

Theorem XII.—*If u is harmonic on E' and satisfies the conditions (B), then*

$$\iint_{E}\left[\left(\frac{\partial u}{\partial x}\right)^2 + \left(\frac{\partial u}{\partial y}\right)^2\right]dxdy = -\int_{\overset{\frown}{C}} u_e\frac{\partial u_e}{\partial n_e}ds.$$

Corollary I.—If we add to the hypotheses of the theorem that $u_e = 0$ *along* C, *then* $u \equiv 0$ *on* E'.

Corollary II.—If we add to the hypotheses of the theorem that $\frac{\partial u_e}{\partial n_e} = 0$ on C, *then* u = *constant on* E'.

IV. Logarithmic Potential of a Double Layer

39. Definition.—We suppose that C

$$C: \qquad x = \xi(s), \quad y = \eta(s), \quad s = 0, \ . \ . \ . \ , l$$

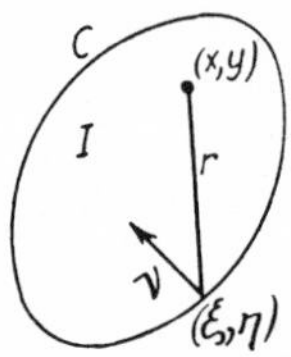

Fig. 11.

has the properties 1) . . . 4) of §36. Let (ξ, η) be a point on C, (x, y) a fixed point not on C, r their distance:

$$r = \sqrt{(x - \xi)^2 + (y - \eta)^2},$$

ν the inner normal to C at (ξ, η), $\mu(s)$ a continuous function on C. Then the definite integral

$$(25) \qquad w(x, y) = \int_0^l \mu(s) \frac{\partial \log \frac{1}{r}}{\partial \nu} ds$$

for physical reasons is called the *logarithmic potential of a double layer* of density μ (s) distributed over the curve C We obtain as follows a more explicit form for $\frac{\partial}{\partial \nu} \log \frac{1}{r}$.

$$(26) \qquad \frac{\partial}{\partial \nu} \log \frac{1}{r} = \cos (\nu\xi) \frac{\partial}{\partial \xi} \log \frac{1}{r} + \cos (\nu\eta) \frac{\partial}{\partial \eta} \log \frac{1}{r}$$

$$(27) \qquad = \frac{1}{r}\left[\frac{x - \xi}{r} \cos (\nu\xi) + \frac{y - \eta}{r} \cos (\nu\eta)\right]$$

$$= \frac{1}{r}\left[\cos (r\xi) \cos (\nu\xi) + \cos (r\eta) \cos (\nu\eta)\right]$$

$$= \frac{\cos (r\nu)}{r}.$$

But from (27), since $\cos(\nu\xi) = -\eta'(s)$, $\cos(\nu\eta) = \xi'(s)$, we have

$$\frac{\partial}{\partial \nu} \log \frac{1}{r} = \frac{\left[y - \eta(s)\right]\xi'(s) - \left[x - \xi(s)\right]\eta'(s)}{r^2}$$

Thus $w(x, y)$ may be written in the form

$$w(x, y) = \int_0^l \mu(s) \frac{\cos(r\nu)}{r} ds,$$

or

$$(28)\quad w(x, y) = \int_0^l \mu(s) \frac{\left[y - \eta(s)\right]\xi'(s) - \left[x - \xi(s)\right]\eta'(s)}{\left[x - \xi(s)\right]^2 + \left[y - \eta(s)\right]^2},$$

which shows explicitly the dependence of the integrand upon the parameters x and y.

40. Properties of w(x, y) at Points Not on C.—Put

$$u(x, y) = \log \frac{1}{r}.$$

Then we have

$$\frac{\partial^2 u}{\partial x^2} = -\frac{1}{r^2} + \frac{2(x - \xi)^2}{r^4}$$
$$\frac{\partial^2 u}{\partial y^2} = -\frac{1}{r^2} + \frac{2(y - \eta)^2}{r^4}.$$

Whence by adding we find $\Delta u = 0$. That is, $\log \frac{1}{r}$ is harmonic. Further

$$\frac{\partial^2}{\partial x^2}\left(\frac{\partial}{\partial \xi} \log \frac{1}{r}\right) = \frac{\partial}{\partial \xi}\left(\frac{\partial^2}{\partial x^2} \log \frac{1}{r}\right)$$

and

$$\frac{\partial^2}{\partial y^2}\left(\frac{\partial}{\partial \xi} \log \frac{1}{r}\right) = \frac{\partial}{\partial \xi}\left(\frac{\partial^2}{\partial y^2} \log \frac{1}{r}\right).$$

Whence, adding, we find

$$\Delta\left(\frac{\partial}{\partial \xi}\log\frac{1}{r}\right) = \frac{\partial}{\partial \xi}\left(\Delta \log\frac{1}{r}\right) = \frac{\partial}{\partial \xi}\Delta u = 0.$$

Therefore $\frac{\partial}{\partial \xi}\log\frac{1}{r}$ is harmonic. Similarly, we can show that $\frac{\partial}{\partial \eta}\log\frac{1}{r}$ is harmonic. Hence it follows from (26) that $\frac{\partial}{\partial \nu}\log\frac{1}{r}$ is harmonic. That is,

$$\Delta\left(\frac{\partial}{\partial \nu}\log\frac{1}{r}\right) = 0. \tag{29}$$

Let us now compute $\Delta w(x, y)$. We have

$$\begin{aligned}\frac{\partial^2}{\partial x^2} w(x, y) &= \frac{\partial^2}{\partial x^2}\int_0^l \mu(s)\frac{\partial}{\partial \nu}\log\frac{1}{r}\,ds \\ &= \int_0^l \mu(s)\frac{\partial^2}{\partial x^2}\left(\frac{\partial}{\partial \nu}\log\frac{1}{r}\right)ds,\end{aligned} \tag{30}$$

since the rules for the differentiation of a definite integral with respect to a parameter are applicable. Likewise, we have

$$\frac{\partial^2}{\partial y^2} w(x, y) = \int_0^l \mu(s)\frac{\partial^2}{\partial y^2}\left(\frac{\partial}{\partial \nu}\log\frac{1}{r}\right)ds. \tag{31}$$

Add (30) and (31) and take account of (29). We find

$$\Delta w(x, y) = \int_0^l \mu(s)\Delta\left(\frac{\partial}{\partial \nu}\log\frac{1}{r}\right)ds = 0.$$

From the explicit expression (28) we see that $w(x, y)$ is single-valued and continuous with all of its derivatives at every point (x, y) not on C. Thus we have proved the following theorem:

Theorem XIII.—*The function*

$$w(x, y) = \int_0^l \mu(s)\frac{\partial}{\partial \nu}\log\frac{1}{r}\,ds$$

is single-valued and continuous with all of its derivatives at every point (x, y) not on C, and in the same domain w is harmonic:

$$\Delta w(x, y) = 0.$$

41. Behavior of w(x, y) on C.—In the preceding discussion (x, y) was supposed *not* to lie on C. Let now (x, y) coincide with a point (x_o, y_o) on C, corresponding to a value $s = s_o$, so that

$$x_o = \xi(s_o), \quad y_o = \eta(s_o).$$

Then the integrand becomes indeterminate. The indeterminate expression which we desire to investigate is

$$\frac{\cos(r_o\nu)}{r_o} = \frac{\left[\eta(s_o) - \eta(s)\right]\xi'(s) - \left[\xi(s_o) - \xi(s)\right]\eta'(s)}{\left[\xi(s_o) - \xi(s)\right]^2 + \left[\eta(s_o) - \eta(s)\right]^2}.$$

To evaluate, apply Taylor's remainder theorem to both numerator and denominator, stopping at the derivatives of the second order. A factor $(s - s_o)^2$ appears in both and cancels. The limit of what remains as $s \rightarrow s_o$ is

$$\frac{\xi'\eta'' - \xi''\eta'}{2(\xi'^2 + \eta'^2)} = \frac{1}{2R_o},$$

where R_o is the radius of curvature of C at (x_o, y_o). $w(x, y)$ is thus defined and has a determinate finite value at every point (x_o, y_o) of C:

$$w(x_o, y_o) =$$

$$\int_0^l \mu(s) \frac{\left[\eta(s_o) - \eta(s)\right]\xi'(s) - \left[\xi(s_o) - \xi(s)\right]\eta'(s)}{\left[\xi(s_o) - \xi(s)\right]^2 + \left[\eta(s_o) - \eta(s)\right]^2} ds.$$

The existence of $w(x, y)$ on the boundary of C does not, however, imply that $w(x, y)$ remains continuous, as (x, y) crosses the boundary C. On the contrary, $w(x, y)$ undergoes a discontinuity as (x, y) crosses C. Let us denote by P_i an interior point of C and by P_e an exterior point (x, y) in the vicinity of $P_o(x_o, y_o)$. If P_i approaches P_o, then $w(x, y)$ approaches a definite finite limit $w_i(x_o, y_o)$. If P_e approaches P_o, then $w(x, y)$ approaches a definite finite limit $w_e(x_o, y_o)$. Between these quantities and $w(x_o, y_o)$ defined above the following relations hold:

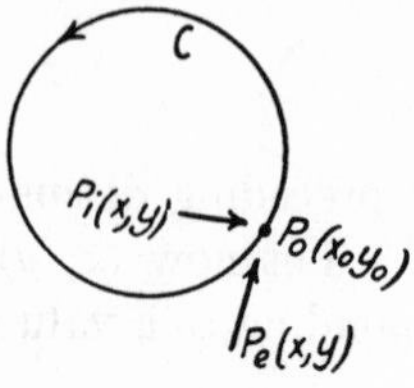

FIG. 12.

$$
\begin{aligned}
w_i(x_o, y_o) &= w(x_o, y_o) + \pi.\mu(s_o) \\
w_e(x_o, y_o) &= w(x_o, y_o) - \pi.\mu(s_o).^1
\end{aligned} \tag{32}
$$

For future use we give here a proof for the special case $\mu(s) \equiv 1$. Let us use the notation

$$\zeta = \xi + i\eta, \quad z = x + iy$$

and consider the integral H along C:

$$H_c = \int_c \frac{d\zeta}{\zeta - z}.$$

By Cauchy's first integral theorem

$$H_c = \begin{cases} 2\pi i, & xy \text{ an interior point of } C. \\ 0, & xy \text{ an exterior point of } C. \end{cases}$$

If we separate the real and the imaginary parts of the integral, we obtain

$$
\begin{aligned}
H_c = &\int \frac{(\xi - x)d\xi + (\eta - y)d\eta}{(x - \xi)^2 + (y - \eta)^2} \\
&+ i\int \frac{(y - \eta)d\xi - (x - \xi)d\eta}{(x - \xi)^2 + (y - \eta)^2} \equiv V_c + iW_c.
\end{aligned}
$$

[1] For a proof of these statements see HORN, "Einfuhrung in die Theorie der Partiellen Differential-Gleichungen," §52.

Hence, by equation (28), $w(x, y)$ is the coefficient of i in the integral H_c. Thus we get

$$w(x, y) = \begin{cases} 2\pi, & xy \text{ an interior point of } C. \\ 0, & xy \text{ an exterior point of } C. \end{cases}$$

But if $w(x, y) = 2\pi$ constantly as x, y varies, then

$$\lim_{P_i \to P_o} w(x, y) = w_i(x_o, y_o) = 2\pi.$$

In like manner

$$\lim_{P_e \to P_o} w(x, y) = w_e(x_o, y_o) = 0.$$

We have now, in order to complete the proof, to compute $w(x_o, y_o)$ for a point (x_o, y_o) *on* C. Draw about P_o an arc α of a circle cutting C in Q and R. This arc will subtend an angle ω at P_o. Designate by C' the path C minus the arc QP_oR. Then $H_{C'} + H_\alpha = 0$, since P_o is an exterior point for this path. But in the elements of the theory of functions of a complex variable it is shown that

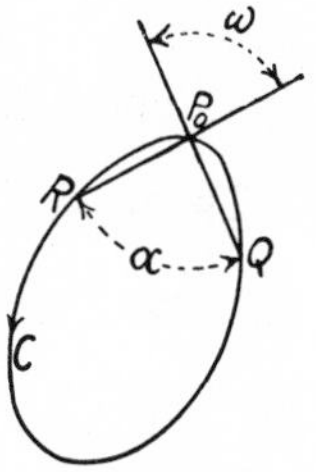

FIG. 13.

$$H_\alpha = \int_\alpha \frac{d\zeta}{\zeta - z_o} = -\varpi i,$$

whence

$$H_{C'} = i\varpi.$$

Therefore,

$$W_{C'} = \varpi.$$

Let the radius of the arc α be ρ. Then, as $\rho \to 0$,

$$W_{C'} \to W_C = w(x_o, y_o).$$

Now the integral W_C is convergent as shown above and $\varpi \to \pi$. Therefore

$$w(x_o, y_o) = \pi.$$

The equations (32) are satisfied for these values of $w(x_o, y_o)$, $w_i(x_o, y_o)$, and $w_e(x_o, y_o)$ for the special case $\mu(s) \equiv 1$.

42. Behavior of $\frac{\partial w}{\partial n}$ on the Boundary C and at Infinity.—At the point $P_o(x_o, y_o)$ draw the interior normal n_i and take on it a point $P_i(x, y)$. From Theorem XIII we know that at P_i, $\frac{\partial w(x, y)}{\partial n_i}$ is definite and finite. Let the point P_i approach P_o along n_i and put

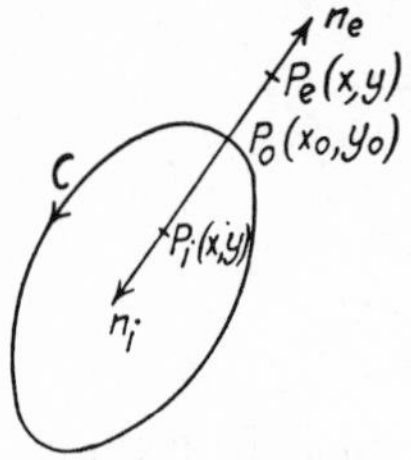

FIG. 14.

$$\lim_{P_i \to P_o} \frac{\partial w}{\partial n_i} = \frac{\partial w_i}{\partial n_i}, \text{ if the limit exists,}$$

$$\lim_{P_e \to P_o} \frac{\partial w}{\partial n_i} = \frac{\partial w_e}{\partial n_i}, \text{ if the limit exists,}$$

P_e designating a point on the prolongation of n_i beyond P_o. Then the following theorem is true.[1]

Theorem XIV.—*If one of the two limits* $\frac{\partial w_i}{\partial n_i}, \frac{\partial w_e}{\partial n_i}$, *exists, then also the other exists and*

$$\frac{\partial w_i}{\partial n_i} = \frac{\partial w_e}{\partial n_i}. \tag{33}$$

If we denote by n_e the exterior normal to C at P_o, then

$$\frac{\partial}{\partial n_i} = -\frac{\partial}{\partial n_e}$$

from the definition of the directional derivative. Hence (33) may also be written

$$\frac{\partial w_i}{\partial n_i} = -\frac{\partial w_e}{\partial n_e}. \tag{34}$$

If we introduce polar coordinates r, θ, then $w(x, y)$ becomes $W(r, \theta)$. Then the following theorem is true:[1]

[1] For proof, see HORN, *Loc. cit.*, §54.

Theorem XV.—

$$\lim_{r \to \infty} W(r, \theta) = 0, \textit{ uniformly as to } \theta.$$

$$\lim_{r \to \infty} rW\frac{\partial W}{\partial r} = 0, \textit{ uniformly as to } \theta.$$

43. Case Where w_i or w_e Vanish Along C.—We can now prove the following theorem:

Theorem XVI.—*If* $w_i(x, y) \equiv 0$ *along* C, *then*

$$\mu(s) \equiv 0.$$

Proof.—Theorem XIII assures us that $w(x, y)$ is harmonic on I'. By hypothesis $w_i(x, y) \equiv 0$ along C. Taking it for granted that the conditions (A) are always satisfied by w, it follows from Corollary I to Theorem XI that

$$w(x, y) \equiv 0 \text{ on } I'.$$

Hence it follows at once that $\frac{\partial w}{\partial n_i} \equiv 0$ for every interior point of C. Therefore, $\frac{\partial w_i}{\partial n_i}$ exists along C and vanishes there. Hence, also by Theorem XIV, $\frac{\partial w_e}{\partial n_e}$ exists along C and vanishes there. By Theorem XV, $\lim_{r \to \infty} rW\frac{\partial W}{\partial r} = 0$, uniformly as to θ. Hence, if we take it for granted that the conditions (B) on w are satisfied, then it follows from Corollary II to Theorem XII that w is constant on E'. We see from Theorem XV that this constant must be zero:

$$w(x, y) \equiv 0 \text{ on } E'.$$

Whence it follows that $w_e(x, y) = 0$ on C. If we now apply (32) we find

$$w_i(x_o, y_o) - w_e(x_o, y_o) = 2\pi\mu(s_o) = 0,$$

uniformly as to s_o. Therefore

$$\mu(s_o) \equiv 0, \text{ uniformly as to } s_o.$$

A similar theorem holds with respect to the limit w_e:

Theorem XVII.—*If* $w_e(x, y) \equiv 0$ *along* C, *then*

$$\mu(s) = \textit{constant}.$$

Proof.—Theorem XIII assures us that w is harmonic on E'. By hypothesis $w_e(x, y) \equiv 0$ along C. Furthermore, $\lim_{r\to\infty} rW\frac{\partial W}{\partial r} = 0$, uniformly as to θ, according to Theorem XV. Taking it for granted that the conditions (B) of §38 are always satisfied by w, it follows from Corollary I to Theorem XII that $w(x, y) \equiv 0$ on E'. Hence it follows at once that $\frac{\partial w}{\partial n_e} \equiv 0$ for every point exterior to C. Therefore $\frac{\partial w_e}{\partial n_e}$ exists along C and vanishes there. Hence also, by Theorem XIV, $\frac{\partial w_i}{\partial n_i}$ exists along C and vanishes there. Taking it for granted that the conditions (A) are always satisfied by w, it follows from Corollary II to Theorem XI that

$$w(x, y) = \text{constant on } I'.$$

Whence it follows that $w_i(x, y) = $ constant on C. If we now apply (32), we find

$$w_i(x_o, y_o) - w_e(x_o, y_o) = 2\pi\mu(s_o) = \text{constant},$$

uniformly as to s_o. Therefore,

$$\mu(s_o) = \text{constant, uniformly as to } s_o.$$

V. Fredholm's Solution of Dirichlet's Problem

44. Dirichlet's Problem.—We formulate Dirichlet's problem as follows:

Given.—1) A closed curve C

$$C: \quad x = \xi(s),\ y = \eta(s),\ 0 \leq s \leq l$$

(35) $$\xi(0) = \xi(l),\ \eta(0) = \eta(l)$$

with the properties 1) . . . 4) of §36.

2) A function $F(s)$, continuous on C for $0 \leq s \leq l$, and

(36) $$F(0) = F(l)$$

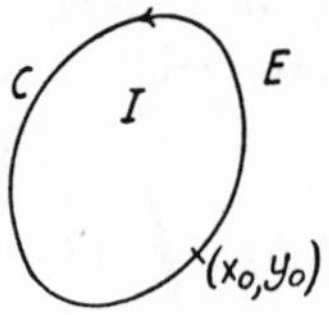

Fig. 15.

Required.—A function $u(x, y)$ such that

α) u is harmonic on I', and

β) $u_i(x_o, y_o) = F(s_o)$, uniformly as to s_o,

where $u_i(x_o, y_o)$ is the limit approached by $u(x, y)$ as the point (x, y) approaches from the interior a point (x_o, y_o) of parameter s_o on the boundary C.

45. Reduction to an Integral Equation. *First Method.*—The function $w(x, y)$:

$$w(x, y) = \int_0^l \mu(s) \frac{\partial}{\partial \nu} \log \frac{1}{r}\, ds$$

is harmonic on I' and thus satisfies the condition (α) for every choice of $\mu(s)$ for which $\mu(s)$ is continuous on $[0l]$ and

$$\mu(0) = \mu(l).$$

The function $w(x, y)$ will then furnish a solution of Dirichlet's problem if $\mu(s)$ can be so determined that the condition (β) is satisfied:

(37) $$w_i(x_o, y_o) = F(s_o), \text{ uniformly as to } s_o.$$

From the first of equations (32) we have

$$w_i(x_o, y_o) = w(x_o, y_o) + \pi\mu(s_o).$$

The substitution of this value of w_i in (37) gives

(38) $$w(x_o, y_o) + \pi\mu(s_o) = F(s_o).$$

If in this equality we substitute for $w(x_o, y_o)$ its explicit expression as given in §41, we obtain

$$(39)\quad \int_0^l \mu(s)\frac{\left[\eta(s_o) - \eta(s)\right]\xi'(s) - \left[\xi(s_o) - \xi(s)\right]\eta'(s)}{\left[\xi(s_o) - \xi(s)\right]^2 + \left[\eta(s_o) - \eta(s)\right]^2}ds + \pi\mu(s_o) = F(s_o).$$

Divide through by π. Then (39) becomes an integral equation for the determination of $\mu(s_o)$ with the kernel

$$(40)\quad K(s_c, s) = \frac{1}{\pi}\frac{\left[\eta(s_o) - \eta(s)\right]\xi'(s) - \left[\xi(s_o) - \xi(s)\right]\eta'(s)}{\left[\xi(s_o) - \xi(s)\right]^2 + \left[\eta(s_o) - \eta(s)\right]^2}.$$

Put $\dfrac{F(s_o)}{\pi} = f(s_o)$. Then the integral equation takes the standard form

$$(41)\qquad \mu(s_o) = f(s_o) - \int_0^l K(s_o, s)\mu(s)ds.$$

This is a special case of the integral equation with a parameter λ, for which $\lambda = -1$.

Thus, in order that $u = w(x, y)$ may be a solution of Dirichlet's problem, it is *necessary* that the density $\mu(s)$ satisfy the integral equation (41). This condition is also *sufficient*. For, suppose $\mu(s)$ to be a continuous solution of (41). Then we have

$K(0, s) = K(l, s)$ on account of (35), and
$f(0) = f(l)$ on account of (36), and hence
$\mu(0) = \mu(l)$ from (41).

The function $w(x, y)$ formed with this function $\mu(s)$ will then satisfy (38) and, therefore, also (37) and, consequently, $u = w(x, y)$ will be a solution of Dirichlet's problem.

Second Method.—From equations (32) we find

$$(42) \qquad \begin{aligned} w_i(x_o, y_o) - w_e(x_o, y_o) &= 2\pi\mu(s_o), \text{ and} \\ w_i(x_o, y_o) + w_e(x_o, y_o) &= 2w(x_o, y_o), \end{aligned}$$

which by §41

$$= 2\int_0^l \mu(s)\,\frac{\cos(r_o\nu)}{r_o}\,ds.$$

We now seek as a solution of Dirichlet's problem a function u which is harmonic everywhere except on C, and which upon C satisfies the condition

$$(43) \qquad u_i + hu_e = F(s_o) + hG(s_o),$$

where h is an arbitrary parameter. For $h = 0$, we have the interior problem, and for $h = \infty$ the exterior problem.

The function $w(x, y)$ is harmonic everywhere except on C. This function will then be a solution of Dirichlet's problem, provided $\mu(s)$ can be so determined that $u = w(x, y)$ satisfies (43). Solve now the equations (42) for $u_i = w_i$ and $u_\epsilon = w_e$ and substitute in (43). We find

$$(44) \qquad \mu(s_o) = f(s_o) + \lambda\int_0^l \mu(s)\frac{\cos(r_o\nu)}{\pi r_o}ds$$

where
$$f(s_o) = \frac{hG(s_o) + F(s_o)}{\pi(1-h)}, \quad \lambda = \frac{h+1}{h-1}.$$

This is an integral equation of the second kind with a parameter λ. The kernel is $\dfrac{\cos(r_o\nu)}{\pi r_o} \equiv K(s_o, s)$ [see (40)]. For $h = 0$, we have the interior problem, but for $h = 0$ we have $\lambda = -1$ and (44) reduces to (41). For $h = \infty$, we have the exterior problem, but for $h = \infty$ we have $\lambda = +1$, and (44) reduces to

$$(45) \qquad \mu(s_o) = \frac{G(s_o)}{\pi} + \int_0^l K(s_o, s)\mu(s)ds.$$

We remark that, if $F(s_o) \equiv 0$ on C and $h = 0$, then (44) becomes a homogeneous linear integral equation with $\lambda = -1$.

46. Solution of the Integral Equation.—The kernel $K(s_o, s)$ is real, continuous, and $\neq 0$ in the region G,

$$G: \qquad 0 \leqq s_o \leqq l, \quad 0 \leqq s \leqq l,$$

and thus the preceding theory of integral equations is applicable. From Fredholm's first fundamental theorem we know that if $D(-1) \neq 0$ then (41) has one and only one continuous solution.

We show first that $D(-1) \neq 0$. For this purpose we use Fredholm's second fundamental theorem, from which it follows that the corresponding homogeneous integral equation

$$\mu(s_o) + \int_0^l K(s_o, s)\mu(s)ds = 0 \tag{46}$$

has no other solution than the trivial one $\mu \equiv 0$, if $D(-1) \neq 0$; while if $\lambda = -1$ is a root of $D(\lambda) = 0$ of index q, then (46) has ∞^q solutions. Hence, if (46) has no other solution than $\mu \equiv 0$, then $D(-1) \neq 0$.

Multiply the members of (46) by π. Then

$$\int_0^l \pi K(s_o, s)\mu(s)ds + \pi\mu(s_o) = 0,$$

uniformly as to s_o. This last equation, by (32), after taking account of our notations in §41 and equations (40), can be written

$$w_i(x_o, y_o) = 0, \text{ uniformly as to } s_o.$$

But, by Theorem XVI, if $w_i(x_o, y_o) = 0$ along C, then

$$\mu(s) \equiv 0.$$

Therefore, (46) has no other continuous solution than $\mu \equiv 0$ and, consequently,

$$D(-1) \neq 0.$$

Thus we see that $\lambda = -1$ is not a characteristic constant for the kernel $K(s_o, s)$. Hence (41) has one and only one continuous solution $\mu(s_o)$ representable by Fredholm's formula

$$\mu(s_o) = f(s_o) + \int_0^l \frac{D(s_o, s; -1)}{D(-1)} f(s)ds.$$

Therefore, *$w(x, y)$ with $\mu = \mu(s_o)$ is a solution of Dirichlet's problem.* We state this result in the following theorem:

Theorem XVIII.—*Given*

1) *A closed curve C*

$$C: \qquad x = \xi(s), \quad y = \eta(s), \quad 0 \leqq s \leqq l$$
$$\xi(0) = \xi(l), \quad \eta(0) = \eta(l)$$

with the properties 1) . . . 4) *of* §36.

2) *A function $F(s)$, continuous on C for $0 \leqq s \leqq l$,*

$$F(0) = F(l).$$

Then there exists a function $u(x, y)$ such that

α) u is harmonic on I', and

β) $u_i(x_o, y_o) = F(s_o)$, uniformly as to s_o, where $u_i(x_o, y_o)$ is the limit approached by $u(x, y)$ as the point (x, y) approaches from the interior a point (x_o, y_o) of parameter s_o on the boundary C. This function is given by

$$u = \int_0^l \mu(s) \frac{\partial}{\partial \nu} \log \frac{1}{r}\, ds,$$

where $\mu(s)$ is the unique solution of (41) *and is given by*

$$\mu(s_o) = f(s_o) + \int_0^l \frac{D(s_o, s; -1)}{D(-1)} f(s)ds,$$

where $\pi f(s_o) = F(s_o)$.

47. Index of $\lambda = 1$ for $K(s_o, s)$.—We have seen that $\lambda = -1$ is not a characteristic constant of $K(s_o, s)$. We shall now prove the following theorem:

Theorem XIX.—$\lambda = +1$ *is a characteristic constant of* $K(s_o, s)$ *of index* 1.

Proof.—According to Fredholm's second fundamental theorem, the statement to be proved is equivalent to the statement that the homogeneous equation

$$\mu(s_o) = \int_0^l K(s_o, s)\mu(s)ds \tag{47}$$

has ∞^1 solutions. Now, if we make use of equations (40) and (39) and the explicit expression for $w(x_o, y_o)$ given in §41, we see that (47) is equivalent to

$$w(x_o, y_o) = \pi\mu(s_o). \tag{48}$$

But from the proof given for the equations (32) for the special case $\mu(s) \equiv 1$, we see that $\mu(s) \equiv 1$ is a solution of (48), for when $\mu(s) \equiv 1$, we have

$$w(x_o, y_o) = \pi = \pi\mu(s_o).$$

This is a solution of (47) which is continuous and not identically zero. Hence $\lambda = 1$ is a *characteristic constant* of $K(s_o, s)$.

We will now determine the index of $\lambda = +1$ for $K(s_o, s)$. To this end we determine all solutions of (47) or of its equivalent (48). Now, on account of the second of equations (32), our equation (48) above reduces to

$$w_e(x_o, y_o) = 0.$$

But, according to Theorem XVII, the most general solution of this equation, and, therefore, of (47), is

$$\mu(s_o) = \text{constant}.$$

Therefore, (47) has ∞^1 solutions, which shows that the index of the characteristic constant $\lambda = +1$ of $K(s_o, s)$ is 1.

VI. Logarithmic Potential of a Simple Layer

48. Definition.—We have given a closed curve C

$$C: \qquad x = \xi(s), \quad y = \eta(s), \quad s = 0, \ . \ . \ . \ , l$$

having the properties 1) . . . 4) of §36. Let r represent the distance between a point (ξ, η) on C and a point (x, y) not on C, $k(s)$ a continuous function on C. Then the definite integral

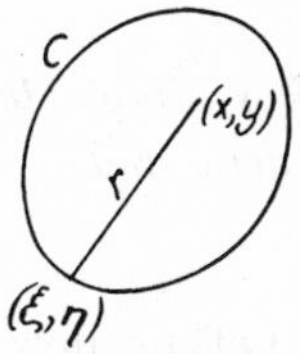

FIG. 16.

$$v(x, y) = \int_0^l k(s) \log \frac{1}{r}\, ds$$
$$= -\frac{1}{2}\int_0^l k(s) \log\left\{\left[x - \xi(s)\right]^2 + \left[y - \eta(s)\right]^2\right\} ds$$

is called the *logarithmic potential of a simple layer* of density $k(s)$ distributed over the curve C.

49. Properties of v(x, y).—

The $\log\left\{\left[x - \xi(s)\right]^2 + \left[y - \eta(s)\right]^2\right\}$ is a regular analytic function of x and y at all points not on C and hence the definite integral as a function of the parameters x and y defines a function $v(x, y)$ which is continuous with all of its derivatives in I' and E'. Further, $v(x, y)$ is harmonic in the same region. For, from the theory of definite integrals

$$\Delta v = \Delta \int_0^l k(s) \log \frac{1}{r} ds$$
$$= \int_0^l k(s) \Delta \log \frac{1}{r}\, ds.$$

But it has been previously shown that

$$\Delta \log \frac{1}{r} = 0.$$

Therefore

$$\Delta v(x, y) = 0.$$

Hence we have proved the following theorem:

Theorem XX.—*The logarithmic potential* $v(x, y)$:

$$v(x, y) = \int_0^l k(s) \log \frac{1}{r}\, ds \tag{49}$$

of a simple layer is continuous with its derivatives of all orders, and is harmonic, on I' *and* E':

$$\Delta v(x, y) = 0.$$

Let us now investigate the behavior of $v(x, y)$ on the boundary C at any point (x_o, y_o) with parameter s_o

$$x_o = \xi(s_o), \quad y_o = \eta(s_o).$$

From the explicit expression for $v(x, y)$:

$$v(x_o, y_o) = -\frac{1}{2}\int_0^l k(s) \times \log \left\{\left[\xi(s_o) - \xi(s)\right]^2 + \left[\eta(s_o) - \eta(s)\right]^2\right\} ds$$

we see that when $s = s_o$, the logarithm which appears in the integrand becomes infinite. Thus we see that we have to do with an improper definite integral. Apply Taylor's remainder theorem to the expression of which the logarithm is taken. We obtain

$$(s - s_o)^2 A,$$

where A does not become zero as s approaches s_o. Then the integrand becomes infinite like $\log (s - s_o)$ as $s \to s_o$. From the theory of improper definite integrals we know that then this definite integral remains definite and finite.

We state without proof[1] the two following theorems, in which $P_i(x, y)$ denotes an interior, $P_e(x, y)$ an exterior point in the vicinity of P_o.

[1] For a proof, see Horn, §53.

Theorem XXI.—*The limit of $v(x, y)$ as $P_i \to P_o$ exists:*

$$\lim_{P_i \to P_o} v(x, y) = v_i(x_o, y_o).$$

The limit of $v(x, y)$ as $P_e \to P_o$ exists:

$$\lim_{P_e \to P_o} v(x, y) = v_e(x, y);$$

and $$v_i(x_o, y_o) = v_e(x_o, y_o) = v(x_o, y_o).$$

That is, $v(x, y)$ remains continuous as the point $P(x, y)$ crosses the boundary C. In this respect the behavior of the function $v(x, y)$ is simpler than that of $w(x, y)$ for a double layer.

Construct for the point $P_o(x_o, y_o)$ the interior normal n_i to C and take on it the interior point P_i and on its exterior prolongation the exterior point P_e; then it follows from Theorem XX that $\frac{\partial v}{\partial n_i}$ exists at P_i and at P_e. For the approach of P_i and P_e toward P_o along the normal the following theorem holds:

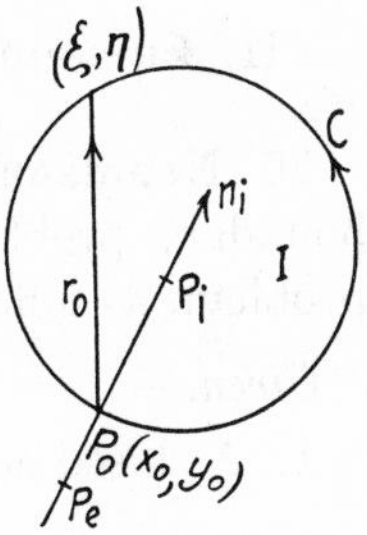

FIG. 17.

Theorem XXII.—*The limit of $\frac{\partial v}{\partial n_i}$ as $P_i \to P_o$ exists:*

$$\lim_{P_i \to P_o} \frac{\partial v}{\partial n_i} = \frac{\partial v_i}{\partial n_i} = -\pi k(s_o) + \int_0^l k(s) \frac{\cos (r_o n_i)}{r_o} ds.$$

The limit of $\frac{\partial v}{\partial n_i}$ as $P_e \to P_o$ exists:

$$\lim_{P_e \to P_o} \frac{\partial v}{\partial n_i} = \frac{\partial v_e}{\partial n_i} = +\pi k(s_o) + \int_0^l k(s) \frac{\cos (r_o n_i)}{r_o} ds,$$

where r_o is the distance from $P_o(x_o, y_o)$ to a variable point $\left[\xi(s), \eta(s)\right]$ of C.

By referring to the figure, we obtain for $\cos(r_o n_i)$ the expression

$$\begin{aligned}\cos(r_o n_i) &= \cos(r_o x)\cos(n_i x) + \cos(r_o y)\cos(n_i y)\\ &= -\frac{\xi(s) - x_o}{r_o}\eta'(s_o) + \frac{\eta(s) - y_o}{r_o}\xi'(s_o),\end{aligned}$$

and hence

$$\frac{\cos(r_o n_i)}{r_o} = \frac{\left[\eta(s) - \eta(s_o)\right]\xi'(s_o) - \left[\xi(s) - \xi(s_o)\right]\eta'(s_o)}{\left[\xi(s) - \xi(s_o)\right]^2 + \left[\eta(s) - \eta(s_o)\right]^2}.$$

VII. Fredholm's Solution of Neumann's Problem

50. Neumann's Problem.—We formulate the second boundary problem of the potential theory, or Neumann's problem, as follows:

Given.—

1) A closed curve C,

$$C: \qquad x = \xi(s), \quad y = \eta(s), \quad 0 \leq s \leq l$$
$$\xi(0) = \xi(l), \quad \eta(0) = \eta(l)$$

with the properties 1) . . . 4) of §36.

2) A function $F(s)$, continuous on C for $0 \leq s \leq l$:

$$F(0) = F(l).$$

Required.—A function $u(x, y)$ such that

α) $\Delta u = 0$ on I'.

β) $\dfrac{\partial u_i}{\partial n_i} = F(s_o)$, uniformly as to s_o.

51. Reduction to an Integral Equation. *First Method.*—The function $v(x, y)$ given by (49) is a function satisfying condition (α) for any choice of $k(s)$ which is continuous on

C. If it is possible to determine the function $k(s)$ so as to satisfy the condition (β):

$$(50) \qquad \frac{\partial v_i}{\partial n_i} = F(s_o), \text{ uniformly as to } s_o,$$

then we will have a solution of our problem.
Equation (50), written explicitly, according to Theorem XXII becomes

$$(51) \quad \int_0^l k(s)\frac{\Big[\eta(s) - \eta(s_o)\Big]\xi'(s_o) - \Big[\xi(s) - \xi(s_o)\Big]\eta'(s_o)}{\Big[\xi(s) - \xi(s_o)\Big]^2 + \Big[\eta(s) - \eta(s_o)\Big]^2} ds$$
$$= F(s_o) + \pi k(s_o).$$

This is an integral equation for the determination of $k(s)$. Divide through by π and put

$$(52) \quad \widetilde{K}(s_o, s) = \frac{1}{\pi}\frac{\Big[\eta(s) - \eta(s_o)\Big]\xi'(s_o) - \Big[\xi(s) - \xi(s_o)\Big]\eta'(s_o)}{\Big[\xi(s) - \xi(s_o)\Big]^2 + \Big[\eta(s) - \eta(s_o)\Big]^2}$$

and
$$f(s_o) = -\frac{F(s_o)}{\pi}.$$

Then (51) becomes

$$(53) \qquad k(s_o) = f(s_o) + \int_0^l \widetilde{K}(s_o, s)k(s)ds.$$

Introduce a parameter λ and write

$$(54) \qquad k(s_o) = f(s_o) + \lambda\int_0^l \widetilde{K}(s_o, s)k(s)ds,$$

which becomes identical with (53) for $\lambda = +1$.

From (51) we see that $k(0) = k(l)$, since, from our hypothesis, $F(0) = F(l)$ and $\widetilde{K}(0, s) = \widetilde{K}(l, s)$.

Second Method.—From Theorem XXII we have

$$(55)\qquad \begin{aligned} \frac{\partial \mathrm{v}_i}{\partial n_i} &= -\pi k(s_o) + \int_0^l k(s)\frac{\cos(r_o n_i)}{r_o}ds \\ \frac{\partial \mathrm{v}_e}{\partial n_i} &= +\pi k(s_o) + \int_0^l k(s)\frac{\cos(r_o n_i)}{r_o}ds. \end{aligned}$$

We now seek, as a solution of Neumann's problem, a function u which is harmonic everywhere except on C, and which on C satisfies the condition

$$(56)\qquad h\frac{\partial u_i}{\partial n_i} + \frac{\partial u_e}{\partial n_i} = hF(s_o) + G(s_o),$$

where h is an arbitrary parameter. For $h = \infty$, we have the so-called interior case, while for $h = 0$ we have the exterior case. This condition is more general than (50) and for $h = \infty$ reduces to (50).

The function $\mathrm{v}(x, y)$ is harmonic everywhere except on C. This function will then be a solution of Neumann's problem provided $k(s)$ can be so determined that $u = \mathrm{v}(x, y)$ satisfies (56). Substitute now from (55) in (56) for

$$\frac{\partial \mathrm{v}_i}{\partial n_i} = \frac{\partial u_i}{\partial n_i} \text{ and } \frac{\partial \mathrm{v}_e}{\partial n_i} = \frac{\partial u_e}{\partial n_i}.$$

We find

$$(57)\qquad k(s_o) = f(s_o) + \lambda \int_0^l k(s)\frac{\cos(r_o n_i)}{\pi r_o}ds,$$

where $\quad f(s_o) = \dfrac{G(s_o) + hF(s_o)}{\pi(1 - h)},\ \lambda = \dfrac{h+1}{h-1}.$

This is an integral equation of the second kind with a parameter λ. The kernel is $\dfrac{\cos(r_o n_i)}{\pi r_o} \equiv \widetilde{K}(s_o, s)$. For $h = 0$, we have the exterior problem, but for $h = 0$ we have $\lambda = -1$. For $h = \infty$, we have the interior problem, but for $h = \infty$ we have $\lambda = +1$ and (57) reduces to (53).

52. Solution of the Integral Equation.—To the equation (54) the Fredholm theory which we have developed can be applied. We look, then, for a solution of (54) when $\lambda = +1$. Now $\lambda = +1$ is a characteristic constant of $\widetilde{K}(s_o, s)$, for $\widetilde{K}(s_o, s)$ can be obtained from the kernel $K(s_o, s)$ for Dirichlet's problem by interchanging s and s_o. Thus $\widetilde{K}$ and K are adjoint kernels. Then from Theorem IX, Chap. III, $\widetilde{K}(s_o, s)$ and $K(s_o, s)$ have the same characteristic constants with the same indices. But $\lambda = +1$ has been shown by Theorem XIX to be a characteristic constant of index 1 for $K(s_o, s)$. Therefore $\lambda = +1$ is a characteristic constant of index 1 for $\widetilde{K}(s_o, s)$. According to Fredholm's third fundamental theorem, (54) has, in general, no solution for $\lambda = +1$. But if $f(s_o)$ satisfies certain $q = 1$ conditions, then there will be a solution. This condition is

$$\int_0^l f(s)\bar{\varphi}(s)ds = 0, \tag{58}$$

where $\bar{\varphi}(s_o)$ is a solution of the associated homogeneous integral equation

$$\bar{\varphi}(s_o) = \int_0^l K(s_o, s)\bar{\varphi}(s)ds, \tag{59}$$

the kernel of which is the adjoint of the kernel of the integral equation (54) under consideration. But $\lambda = +1$ is a characteristic constant of $K(s_o, s)$ of index 1 and $\bar{\varphi}(s) \equiv 1$ is a solution of (59), as shown in §47. Therefore, condition (58) becomes

$$\int_0^l f(s)ds = 0. \tag{60}$$

Unless this condition is satisfied, there is no solution. If this condition is satisfied, then there are ∞^1 solutions. By referring to the definition of $f(s)$, we see that (60) gives the condition

$$\int_0^l F(s)ds = 0.$$

Thus we see that $\mathrm{v}(x, y)$ is a solution of Neumann's problem. Furthermore, by referring to Theorems XX, XXI, and XXII, we see that $\mathrm{v}(x, y)$ satisfies the conditions (A) of §38. Let $\bar{\mathrm{v}}(x, y)$ be any other solution of Neumann's problem which satisfies the conditions (A). Form the difference

$$\mathfrak{w}(x, y) = \bar{\mathrm{v}}(x, y) - \mathrm{v}(x, y).$$

Then it follows that

1) $\Delta\mathfrak{w} = 0$ on I'.

2) $\dfrac{\partial \mathfrak{w}_i}{\partial n_i} = 0$ along C.

From Corollary II to Theorem XI it follows that

$$\mathfrak{w} = C \text{ (a constant) on } I'$$

Therefore,

$$\bar{\mathrm{v}} = \mathrm{v} + c.$$

Hence, there are ∞^1 solutions of Neumann's problem which differ only by an additive constant. Furthermore, these are the only solutions of the problem which satisfy the conditions (A). We have now the following theorem:

Theorem XXIII.—*Given a closed curve C satisfying the conditions* 1) . . . 4) *of* §36

$$C: \qquad x = \xi(s), \quad y = \eta(s), \quad 0 \leqq s \leqq l$$
$$\xi(0) = \xi(l), \; \eta(0) = \eta(l),$$

and the function $F(s)$ continuous on C, $F(0) = F(l)$, and

$$\int_0^l F(s)ds = 0, \tag{61}$$

then Neumann's problem has ∞^1 solutions

$$\bar{u} = u + c$$

where

$$u = \int_0^l k(s) \log \frac{1}{r}\, ds$$

and $k(s)$ *is a solution of*

$$k(s_o) = f(s_o) + \lambda \int_0^l \widetilde{K}(s_o, s) k(s) ds$$

for $\lambda = 1$, *where* $\widetilde{K}(s_o, s)$ *is defined by* (52). *If the condition* (61) *is not satisfied, the problem has no solution.*

CHAPTER V

HILBERT-SCHMIDT THEORY OF INTEGRAL EQUATIONS WITH SYMMETRIC KERNELS

Solution Expressed in Terms of a Set of Fundamental Functions

I. Existence of at Least One Characteristic Constant

53. Introductory Remarks.—The Fredholm theory of a linear integral equation with a parameter λ:

$$(1) \qquad u(x) = f(x) + \lambda \int_a^b K(x, t)u(t)dt$$

has been developed under the assumptions

A) $K(x, t)$ real.
B) $K(x, t)$ continuous.
C) $K(x, t) \not\equiv 0$ on $R : a \leqq x \leqq b,\ a \leqq t \leqq b$.

For the Hilbert-Schmidt theory a fourth assumption is made that the kernel is symmetric.

(2) *D*) $K(x, t) = K(t, x)$.

It is clear that the results of the Fredholm theory still hold. But, besides, a number of new results will follow from the additional condition *D*). These results were first obtained by Hilbert.

Fredholm obtained the solution of (1) by considering it as the limit of a set of linear equations. He did not carry through the limiting processes, but guessed at the solution and then showed independently of the linear equations that the guess was correct. Hilbert started from the finite system of linear equations, of which equations (1) may be considered as a limit, and actually carried through the limiting process in detail. E. Schmidt, in 1906, obtained Hilbert's results directly without using the limiting process. Hilbert and Schmidt do not make use of Fredholm's results. We shall, however, avail ourselves of these results whenever they lead to simplifications of the proofs.

The first fundamental theorem of Hilbert's theory is the following:

Theorem I.—*Every symmetric kernel has at least one characteristic constant (real or imaginary).*

Our proof of this theorem involves several lemmas which we will give first.

We give an example due to Kowalewski, which illustrates that a kernel which is not symmetric does not necessarily have a characteristic constant. The kernel

$$K(x, t) = \sin(\pi x) \cos(\pi t), \quad [01]$$

has no characteristic constants, if

$$u(x) = \lambda \int_0^1 \sin(\pi x) \cos(\pi t) u(t) dt \tag{3}$$

has no solution other than $u \equiv 0$.

Equation (3) may be written

$$u(x) = \lambda \sin \pi x \int_0^1 \cos(\pi t) u(t) dt \equiv c \sin \pi x,$$

where c is a constant yet to be determined. Substitute this expression for $u(x)$ in (3). We obtain

$$c \sin \pi x = c\lambda \sin \pi x \int_0^1 \cos(\pi t) \sin(\pi t) dt.$$

But
$$\int_0^1 \cos(\pi t) \sin(\pi t)\, dt = 0.$$

Therefore
$$c \sin \pi x \equiv 0,$$
whence
$$c = 0,$$
and
$$u \equiv 0.$$

Thus this particular kernel has no characteristic constants. This shows that Theorem I indeed states a property peculiar to *symmetric* kernels.

54. Power Series for $\frac{D'(\lambda)}{D(\lambda)}$.—The following proof of the theorem is due to Kneser. It is based upon a lemma concerning the expansion of $\frac{D'(\lambda)}{D(\lambda)}$, which holds under the assumptions *A*), *B*), *C*).

From Fredholm's first fundamental theorem, if $D(\lambda) \neq 0$, then (1) has one and only one solution given by

$$u(x) = f(x) + \int_a^b \frac{D(x, t; \lambda)}{D(\lambda)} f(t)dt. \tag{4}$$

Consider the expansion of $\frac{D(x, t; \lambda)}{D(\lambda)}$ as a power series in λ. Now $D(0) = 1 \neq 0$ and, therefore, there exists a ρ such that if $|\lambda| < \rho$, then

$$D(\lambda) \neq 0, \quad |\lambda| < \rho. \tag{5}$$

Hence, since $D(\lambda)$ is permanently convergent, it can be expanded[1] into a power series convergent for $|\lambda| < \rho$. Then

$$\frac{1}{D(\lambda)} = d_0 + d_1\lambda + d_2\lambda^2 + \cdots, \quad |\lambda| < \rho. \tag{6}$$

Now, $D(x, t; \lambda)$ is a permanently converging power series in λ. Hence, also, the product $D(x, t; \lambda)\cdot\frac{1}{D(\lambda)}$ can be expanded into a power series in λ:

$$\frac{D(x, t; \lambda)}{D(\lambda)} = \sum_{n=1}^{\infty} g_n(x, t)\lambda^n \tag{7}$$

[1] HARKNESS and MORLEY, "Theory of Functions," §83.

and the series will be uniformly convergent as to x and t in R for $|\lambda| < \rho$. The right-hand member of (7) has no term free from λ for

$$D(x, t; \lambda) = \lambda K(x, t) + \cdots$$

Substitute from (7) in (4) and we obtain

$$(8)\qquad u(x) = f(x) + \sum_{n=1}^{\infty} \lambda^n \int_a^b g_n(x, t) f(t) dt, \; |\lambda| < \rho.$$

We have been able to interchange the order of the integration and summation in (8) on account of the uniform convergence of $\sum_{n=1}^{\infty} g_n(x, t)\lambda^n$.

The coefficients $g_n(x, t)$ can be determined by comparing (8) with the expression for the solution obtained by successive approximations (see §8):

$$(9)\qquad u(x) = f(x) + \sum_{n=1}^{\infty} \lambda^n \int_a^b K_n(x, t) f(t) dt$$

where $K_n(x, t)$ are the iterated kernels given by

$$(10)\qquad K_n(x, t) = \int_a^b K(x, s) K_{n-1}(s, t) ds.$$

This solution by successive approximations is valid if $|\lambda| < \frac{1}{M(b-a)}$, where M is the maximum of $|K(x, t)|$ on R. Denote by r the smaller of the two quantities ρ, $\frac{1}{M(b-a)}$. Then (8) and (9) hold simultaneously for $|\lambda| < r$. But (1) has one and only one solution; therefore (8) and (9) represent the same function, and coefficients of corresponding powers of λ must be equal. Therefore,

$$\int_a^b g_n(x, t) f(t) dt = \int_a^b K_n(x, t) f(t) dt, \; (x)$$

where the notation (x) is used to mean uniformly as to x. Hence

$$(11) \qquad \int_a^b \left\{ K_n(x, t) - g_n(x, t) \right\} f(t)dt = 0, \ (x).$$

But $K_n(x, t)$ and $g_n(x, t)$ are independent of f and, for a given value of x,

$$K_n(x, t) - g_n(x, t) = M(t)$$

is a real and continuous function of t. Now (11) holds for any choice of f which is continuous. Choose then $f(t) = M(t)$. Then (11) becomes

$$\int_a^b \left[M(t) \right]^2 dt = 0.$$

Hence

$$M(t) \equiv 0 \text{ on } \left[ab \right].$$

Therefore

$$g_n(x, t) \equiv K_n(x, t), \quad (x, t)$$

and (7) becomes

$$(12) \qquad \frac{D(x, t; \lambda)}{D(\lambda)} = \sum_{n=1}^{\infty} K_n(x, t)\lambda^n, \ |\lambda| < r.$$

Thus, under the assumptions A), B), C), we have proved the following theorem:

Theorem II.—*For all sufficiently small values of* λ

$$\frac{D(x, t; \lambda)}{D(\lambda)} = \sum_{n=1}^{\infty} K_n(x, t)\lambda^n,$$

the series converging uniformly as to x *and* t *in* R.

Theorem II holds when $t = x$. Then

$$\frac{D(x, x; \lambda)}{D(\lambda)} = \sum_{n=1} K_n(x, x)\lambda^n, \ |\lambda| < r$$

and $\sum K_n(x, x)\lambda^n$ is uniformly convergent on $\left[ab \right]$

Integrate this equality with respect to x from a to b. We obtain

$$\frac{1}{D(\lambda)}\int_a^b D(x, x; \lambda)dx = \sum_{n=1}^{\infty}\lambda^n \int_a^b K_n(x, x)dx.$$

Let us introduce the permanent notation

$$\int_a^b K_n(x, x)dx = U_n, \text{ a constant,} \tag{13}$$

and recall from Chap. III, §20, equation (27) that

$$\int_a^b D(x, x; \lambda) = -\lambda D'(\lambda).$$

Then our equality above becomes

$$-\frac{\lambda D'(\lambda)}{D(\lambda)} = \sum_{n=1}^{\infty} U_n\lambda^n,$$

which may be written

$$\frac{D'(\lambda)}{D(\lambda)} = -\sum_{n=0}^{\infty} U_{n+1}\lambda^n, \ |\lambda|<r.$$

We have thus the following corollary to Theorem II:

Corollary.—

$$\frac{D'(\lambda)}{D(\lambda)} = -\sum_{n=0}^{\infty} U_{n+1}\lambda^n, \ |\lambda|<r. \tag{14}$$

55. Plan of Kneser's Proof.—Suppose that $K(x, t)$ has no characteristic constant; that is, that $D(\lambda) = 0$ has no roots, real or imaginary. Then the quotient $\frac{D'(\lambda)}{D(\lambda)}$ can be directly expanded into a power series

$$\frac{D'(\lambda)}{D(\lambda)} = \sum_{n=0}^{\infty} C_n\lambda^n,$$

and this series will be permanently converging.[1] But (14) holds for $|\lambda| < r$. Then

$$\sum_{n=0}^{\infty} C_n\lambda^n = \sum_{n=0}^{\infty} U_{n+1}\lambda^n, \ |\lambda| < r,$$

hence

$$C_n = U_{n+1}$$

and, therefore,

$$\sum_{n=0}^{\infty} U_{n+1}\lambda^n \text{ is permanently convergent.}$$

Then also[2] is

$$\sum_{n=0}^{\infty} |U_{n+1}||\lambda|^n \text{ permanently convergent,}$$

and also

$$(15) \qquad \sum_{n=0}^{\infty} |U_{n+1}||\lambda|^{n+1} \text{ is permanently convergent.}$$

Then the series formed by omitting any number of terms from (15) is also permanently convergent. Hence

$$(16) \qquad \sum_{n=0}^{\infty} |U_{2n}||\lambda|^{2n} \text{ is permanently convergent.}$$

We have proved that (16) was permanently convergent by assuming that $K(x, t)$ had no characteristic constant. If, then, for a given kernel $K(x, t)$ we can show that the corresponding series (16) is *not* permanently convergent, we will have shown that $K(x, t)$ has at least one characteristic constant. We are then going to prove that if, in addition to the properties $A)$ $B)$ $C)$, the kernel $K(x, t)$ is symmetric, then the corresponding series (16) is not permanently convergent. Theorem I will then be proved.

[1] Harkness and Morley, *Loc. cit.*, §83.

[2] Harkness and Morley, *Loc. cit.*, §76.

For this purpose we need some auxiliary lemmas on iterated kernels.

56. Lemmas on Iterations of a Symmetric Kernel.—The iterated kernel $K_n(x, t)$ was defined by

$$(17)\qquad \begin{aligned} K_1(x, t) &= K(x, t) \\ K_n(x, t) &= \int_a^b K(x, s)K_{n-1}(s, t)ds. \end{aligned}$$

By successive applications of this recursion formula we find

$$(18)\quad K_n(x, t) = \int_a^b \cdots \int_a^b K(x, s_1)K(s_1, s_2) \cdots K(s_{n-1}, t)ds_1 \cdots ds_{n-1}.$$

In Chap. II, §11, we showed that

$$(19)\qquad K_{n+p}(x, t) = \int_a^b K_n(x, s)K_p(s, t)ds.$$

We now establish the following lemmas:

Lemma I.—If $K(x, t)$ is symmetric, then $K_n(x, t)$ is symmetric.

Proof.—By (18) we have

$$K_n(t, x) = \int_a^b \cdots \int_a^b K(t, s_1)K(s_1, s_2) \cdots K(s_{n-1}, x)\, ds_1 \cdots ds_{n-1}$$

$$= \int_a^b \cdots \int_a^b K(s_1, t)K(s_2, s_1) \cdots K(x, s_{n-1})\, ds_1 \cdots ds_{n-1}$$

on account of the symmetry of $K(x, t)$

$$= \int_a^b \cdots \int_a^b K(x, s_{n-1}) \cdots K(s_2, s_1)K(s_1, t)\, ds_1 \cdots ds_{n-1}.$$

Now make a change of notation.

For	s_1	s_2	$\cdots$	s_{n-2}	s_{n-1}
put	s_{n-1}	s_{n-2}	$\cdots$	s_2	s_1.

Then

$$K_n(t, x) = \int_a \cdots \int_a^b K(x, s_1)K(s_1, s_2) \cdots K(s_{n-1}, t)\, ds_1 \cdots ds_{n-1}$$

$$= K_n(x, t) \text{ by (18).}$$

Lemma II.—If $K(x, t)$ satisfies the conditions A) B) C) D), then $K_n(x, t) \not\equiv 0$ on R.

Proof.—Suppose $K_n(x, t) \equiv 0$ on R, and suppose that n is the lowest index for which an iterated kernel vanishes identically:

$$K(x, t) \not\equiv 0, K_2(x, t) \not\equiv 0, \ldots, K_{n-1}(x,t) \not\equiv 0,$$
$$K_n(x, t) \equiv 0, R.$$

Then certainly, by C), we have $n > 1$. Then from (19) it follows that all following iterated kernels are identically zero. Then

$$K_{n+1}(x, t) \equiv 0 \text{ on } R.$$

Now either n or $n + 1$ must be an even number $2m$:

$$2m = n, \text{ or } n + 1.$$

Then

$$K_{2m}(x, t) \equiv 0 \text{ on } R.$$

But by (19)

$$K_{2m}(x, t) = \int_a^b K_m(x, s)K_m(s, t)ds.$$

This equality holds when $t = x$. Then

$$K_{2m}(x, x) = \int_a^b K_m(x, s)K_m(s, x)ds$$

(20) $$= \int_a^b \left[K_m(x, s) \right]^2 ds$$

on account of the symmetry of $K(x, t)$.

But $K_{2m}(x, t) \equiv 0$ on R. Therefore, since $K_m(x, t)$ is a real function of x and t, we have

$$K_m(x, t) \equiv 0 \text{ on } R.$$

But $$m = \frac{n}{2}, \text{ or } \frac{n+1}{2}.$$

Hence $$n - m = \frac{n}{2}, \text{ or } \frac{n-1}{2} > 0 \text{ for } n > 1.$$

Therefore $$m > n.$$

Thus we have contradicted the assumption that $K_n(x, t)$ was the first iterated function to vanish and the lemma is proved.

It is now clear that if $K(x, t)$ satisfies the conditions $A)$ $B)$ $C)$ $D)$, then $K_n(x, t)$ satisfies the same conditions.

Corollary.— $$U_{2m} > 0.$$

Proof.— $$U_{2m} = \int_a^b K_{2m}(x, x)dx.$$

Now $K_{2m}(x, x) > 0$ by (20), since $K_m(x, t) \not\equiv 0$.

57. Schwarz's Inequality.—Let $\varphi(x)$ and $\psi(x)$ be real and continuous on the interval $[ab]$, and u and v any two reals constant with respect to x. Then

$$\int_a^b \left[u\varphi(x) + \mathrm{v}\psi(x) \right]^2 dx \geqq 0,$$

since we have the square of a real function.

Expand and we obtain

$$u^2 \int_a^b \varphi^2(x)dx + 2u\mathrm{v} \int_a^b \varphi(x)\psi(x)dx + \mathrm{v}^2 \int_a^b \psi^2(x)dx \geqq 0.$$

This is a definite quadratic form and hence

$$(21) \quad \left[\int_a^b \varphi(x)\psi(x)dx \right]^2 \leqq \int_a^b \varphi^2(x)dx \times \int_a^b \psi^2(x)dx.$$

This is called the inequality of Schwarz. In like manner, we can prove that if $\varphi(x, y)$ and $\psi(x, y)$ are real and continuous on R in x and y, then

$$\left[\int_a^b \int_a^b \varphi(x, y)\psi(x, y)dxdy \right]^2$$
$$\leqq \left[\int_a^b \int_a^b \varphi^2(x, y)dxdy \right] \left[\int_a^b \int_a^b \psi^2(x, y)dxdy \right].$$

58. Application of Schwarz's Inequality.—We have obtained one expression for U_{2n}:

$$(13) \qquad U_{2n} = \int_a^b K_{2n}(x, x)dx$$

$$(22) \qquad = \int_a^b \int_a^b \Big[K_n(x, t) \Big]^2 dtdx > 0 \text{ by Lemma II.}$$

We now proceed to determine a second expression for U_{2n}. Now by (19) we have

$$K_{2n}(x, t) = \int_a^b K_{n-1}(x, s)K_{n+1}(s, t)dt.$$

If we put $t = x$, this equality becomes

$$K_{2n}(x, x) = \int_a^b K_{n-1}(x, t)K_{n+1}(t, x)dt$$
$$= \int_a^b K_{n-1}(x, t)K_{n+1}(x, t)dt,$$

on account of the symmetry of $K_{n+1}(x, t)$. Then

$$(23) \qquad U_{2n} = \int_a^b \int_a^b K_{n-1}(x, t)K_{n+1}(x, t)dtdx.$$

Apply Schwarz's inequality to the right member of (23). We obtain

$$\left[\int_a^b \int_a^b K_{n-1}(x, t)K_{n+1}(x, t)dtdx\right]^2$$
$$\leqq \left[\int_a^b \int_a^b K^2_{n-1}(x, t)dtdx\right]\left[\int_a^b \int_a^b K^2_{n+1}(x, t)dtdx\right],$$

which, if we make use of (22) and (23), may be written

$$U^2_{2n} \leqq U_{2n-2} \cdot U_{2n+2}.$$

Divide both members of this inequality by $U_{2n-2} \cdot U_{2n}$. This is possible, since $U_{2n-2} \neq 0$, $U_{2n} \neq 0$. We obtain

$$(24) \qquad \frac{U_{2n+2}}{U_{2n}} \geqq \frac{U_{2n}}{U_{2n-2}}.$$

Putting successively $n = 2, 3, \ldots, n$, we obtain in this way the sequence of inequalities

$$\frac{U_{2n+2}}{U_{2n}} \geq \frac{U_{2n}}{U_{2n-2}} \geq \cdots \geq \frac{U_6}{U_4} \geq \frac{U_4}{U_2}.$$

Therefore

$$(25) \qquad \frac{U_{2n+2}}{U_{2n}} \geq \frac{U_4}{U_2}, \ (n).$$

Apply now the ratio test to the series

$$(16) \qquad \sum_{n=1}^{\infty} U_{2n}|\lambda|^{2n} \equiv \sum_{n=1}^{\infty} V_n$$

to find the radius of convergence in λ. We find

$$\frac{V_{n+1}}{V_n} = \frac{U_{2n+2}}{U_{2n}} |\lambda|^2 \geq \frac{U_4}{U_2} |\lambda|^2, \ (n).$$

Therefore, the series diverges if

$$\frac{U_4}{U_2} |\lambda|^2 > 1,$$

that is, if

$$|\lambda| > \sqrt{\frac{U_2}{U_4}}.$$

Therefore, the series is not a permanently convergent power series in λ. This completes the proof of Theorem I.

Theorem I.—*For a real, symmetric, continuous, non-identically, vanishing kernel $K(x, t)$ there exists at least one characteristic constant λ_o.*

We remark that for every value of λ in the λ-plane, without a circle C of radius $\sqrt{\frac{U_2}{U_4}}$, $\sum_{n=1}^{\infty} U_{2n}|\lambda|^{2n}$ diverges. Hence, from (14) and (16), $\frac{D'(\lambda)}{D(\lambda)}$, when expressed as a series in λ, diverges for all values of λ without the circle C. Then, certainly, $D(\lambda) = 0$ has at least one root within the interior or on the boundary of C

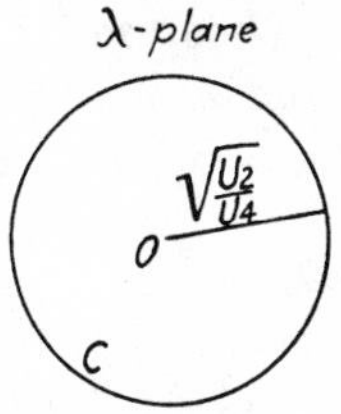

FIG. 18.

II. Orthogonality

59. Orthogonality Theorem.—From the Fredholm theory 1) Two associated kernels $\widetilde{K}(x, t)$, $K(x, t)$:

$$\widetilde{K}(x, t) = K(t, x)$$

have the same characteristic constants with the same indices; if, then

2) λ_o and λ_1 ($\lambda_o \neq \lambda_1$) are two characteristic constants of $K(x, t)$, and hence of $K(t, x)$, and $\varphi_o(x)$ is a fundamental function of $K(x, t)$ for λ_o, and $\tilde{\varphi}_1(x)$ is a fundamental function of $\widetilde{K}(x, t)$ for λ_1, it follows that $\int_a^b \varphi_o(x)\tilde{\varphi}_1(x)dx = 0$.

But if the kernel $K(x, t)$ is symmetric, then $\tilde{\varphi}_1(x)$ is also a fundamental function of $K(x, t)$ for λ_1. If, then, we write $\tilde{\varphi}_1 \equiv \varphi_1(x)$, we have

$$\int_a^b \varphi_o(x)\varphi_1(x)dx = 0.$$

Definition.—Two functions $\varphi(x)$, $\psi(x)$, continuous on the interval $[ab]$ are said to be orthogonal on $[ab]$ if

$$(\varphi\psi) = \int_a^b \varphi(x)\psi(x)dx = 0.$$

We have thus proved the following theorem which is called the *orthogonality theorem:*

Theorem III.—*If $K(x, t)$ is symmetric and $\varphi_o(x)$, $\varphi_1(x)$ are fundamental functions of $K(x, t)$ for λ_o and λ_1 respectively ($\lambda_o \neq \lambda_1$), then $\varphi_o(x)$ and $\varphi_1(x)$ are orthogonal on the interval $[ab]$.*

$$\int_a^b \varphi_o(x)\varphi_1(x)dx = 0. \tag{26}$$

As an illustration of Theorem III, take the symmetric kernel $K(x, t)$:

$$K(x, t) = \begin{cases} (1 - t)x, & [0t] \\ (1 - x)t, & [t1] \end{cases},$$

which has the characteristic constants

$$\lambda_n = n^2\pi^2, \quad n = 1, 2, \ldots$$

and the corresponding fundamental functions

$$\varphi_n(x) = \sin n\pi x.$$

It is then well known from the theory of definite integrals that

$$\int_0^1 \sin n\pi x \sin p\pi x dx = 0, \quad (n \neq p). \tag{27}$$

60. Reality of the Characteristic Constants.—By means of the preceding theorem we can now prove the following theorem due to Hilbert:

Theorem IV.—*If $K(x, t)$ is real and symmetric, continuous, and $\not\equiv 0$, then all of the characteristic constants are real.*

Proof.—Suppose that there is a characteristic constant λ_o not real:

$$\lambda_o = \mu_o + i\nu_o, \quad \nu_o \neq 0.$$

Then the homogeneous equation

$$u(x) = \lambda_o \int_a^b K(x, t)u(t)dt$$

has at least one continuous solution $\varphi(x) \not\equiv 0$:

$$\varphi(x) = (\mu_o + i\nu_o) \int_a^b K(x, t)\varphi(t)dt. \tag{28}$$

1) Suppose $\varphi(x)$ real, then separating the real and imaginary parts of (28), we obtain

$$\varphi(x) = \mu_o \int_a^b K(x, t)\varphi(t)dt \tag{29}$$

$$0 = \nu_o \int_a^b K(x, t)\varphi(t)dt. \tag{30}$$

From (30)

$$\int_a^b K(x, t)\varphi(t)dt = 0,$$

since $\nu_o \neq 0$. Then from (29), $\varphi(x) \equiv 0$, which contradicts the assumption that $\varphi(x) \not\equiv 0$. Therefore, $\varphi(x)$ cannot be real.

2) Suppose $\varphi(x) = \mathrm{v}(x) + iw(x)$, where $\mathrm{v}(x)$ and $w(x)$ are real. Then (28) becomes

$$(31)\quad \mathrm{v}(x) + iw(x) = (\mu_o + i\nu_o)\int_a^b K(x, t)\Big[\mathrm{v}(t) + iw(t)\Big]dt.$$

Separate (31) into its real and imaginary parts. We obtain

$$(32)\qquad \mathrm{v}(x) = \mu_o\int_a^b K(x, t)\mathrm{v}(t)dt - \nu_o\int_a^b K(x, t)w(t)dt$$

$$(33)\quad w(x) = \mu_o\int_a^b K(x, t)w(t)dt + \nu_o\int_a^b K(x, t)\mathrm{v}(t)dt.$$

Multiply both sides of (33) by $-i$ and add to (32). We obtain

$$\mathrm{v}(x) - iw(x) = (\mu_o - i\nu_o)\int_a^b K(x, t)\Big[\mathrm{v}(t) - iw(t)\Big]dt.$$

Therefore $\mathrm{v}(x) - iw(x) \equiv \bar{\varphi}(x)$ is also a fundamental function of $K(x, t)$, belonging to $\bar{\lambda}_0 \equiv \mu_o - i\nu_o$. If we now apply the orthogonality Theorem III, we obtain, since $\bar{\lambda}_0 \neq \lambda_0$,

$$\int_a^b \varphi(x)\bar{\varphi}(x)dx = 0,$$

which may be written

$$\int_a^b\Big[v^2(x) + w^2(x)\Big]dx = 0.$$

But $v(x)$ and $w(x)$ are real functions. Therefore

$$v^2(x) + w^2(x) = 0,\ (x)$$

and hence $\qquad v(x) \equiv 0,\ w(x) \equiv 0$

and therefore $\qquad \varphi(x) \equiv 0,$

which constitutes a contradiction. Therefore λ_o cannot be of the form $\lambda_o = \mu_o + i\nu_o(\nu_o \neq 0)$ and hence λ_o *must be real.*

This result might have been foreseen from the analogy with the finite system of linear equations:

$$u(t_i) - \lambda h \sum_{j=1}^{n} K(t_i, t_j)u(t_j) = f(t_i) \quad (i = 1, \ldots, n)$$

with determinant

$$(34)\quad \Delta = \begin{vmatrix} 1 - \lambda h K_{11} & -\lambda h K_{12} & \cdots & -\lambda h K_{1n} \\ -\lambda h K_{21} & 1 - \lambda h K_{22} & \cdots & -\lambda h K_{2n} \\ \cdots & \cdots & \cdots & \cdots \\ \cdots & \cdots & \cdots & \cdots \\ -\lambda h K_{n1} & -\lambda h K_{n2} & \cdots & 1 - \lambda h K_{nn} \end{vmatrix}.$$

But it is well known[1] that, if K_{ij} is real and $K_{ij} = K_{ji}$, then all of the roots of (34) are real.

In the Fredholm theory we proved that $q \leqq r$ where q is the index of λ_o and r is the multiplicity of the root λ_o of $D(\lambda_o) = 0$. For the case of a symmetric kernel, Hilbert has proved[2] the following more definite theorem which we state here without proof:

Theorem V.—*For a real symmetric kernel the index q of a characteristic constant λ_o is always equal to its multiplicity r: $q = r$.*

61. Complete Normalized Orthogonal System of Fundamental Functions.—Let us suppose that λ_o is a characteristic constant of the symmetric kernel $K(x, t)$ of index q. Then the equation

$$u(x) = \lambda_o \int_a^b K(x, t)u(t)dt$$

has been shown to have q linearly independent solutions:

$$\varphi_\alpha(x) = \frac{D\begin{pmatrix} x_1', & \ldots, & x'_{\alpha-1}, & x\ , & x'_{\alpha+1}, & \ldots, & x_q' \\ y_1', & \ldots, & y'_{\alpha-1}, & y_\alpha', & y'_{\alpha+1}, & \ldots, & y_q' \end{pmatrix}\lambda_o}{D\begin{pmatrix} x_1', & \ldots, & x'_{\alpha-1}, & x_\alpha', & x'_{\alpha+1}, & \ldots, & x_q' \\ y_1', & \ldots, & y'_{\alpha-1}, & y_\alpha', & y'_{\alpha+1}, & \ldots, & y_q' \end{pmatrix}\lambda_o}$$

$$\alpha = 1, \ldots, q.$$

[1] See Bocher, "Introduction to Higher Algebra," §59.

[2] For proof, see Horn, §39.

For a real symmetric kernel, λ_o and, therefore, also $\varphi_\alpha(x)$ are real, since the arguments in the numerator and denominator of the expression for $\varphi_\alpha(x)$ are all real. Any other solution $\psi_\alpha(x)$ is expressible linearly in terms of these q solutions, or in terms of q linearly independent linear combinations of the $\varphi_\alpha(x)$'s:

$$\psi_\alpha(x) = c_{\alpha_1}\varphi_1(x) + \cdot\cdot\cdot + c_{\alpha q}\varphi_q(x), \tag{35}$$

where the determinant of the coefficients $|c_{\alpha\beta}| \neq 0$, $\left[\alpha, \beta = 1, \ldots, q\right]$ does not vanish.

We are going to choose such a system of fundamental functions in such a way that they form what is called a *normalized orthogonal system.*

a) Normalized Fundamental Functions.—A function ψ is said to be *normalized* if

$$(\psi\psi) \equiv \int_a^b \psi^2(x)dx = 1.$$

If $\varphi(x)$ is a fundamental function belonging to λ_o:

$$\varphi(x) = \lambda_o \int_a^b K(x, t)\varphi(t)dt, \quad \left[\varphi(x) \not\equiv 0\right]$$

then $\psi(x) = c\varphi(x)$ $(c \neq 0)$ is clearly again a fundamental function belonging to λ_o, and we can choose c so that

$$(\psi\psi) = 1, \tag{36}$$

that is

$$c^2 \int_a^b \varphi^2(x)dx = 1,$$

whence

$$c = \pm \frac{1}{\sqrt{\int_a^b \varphi^2(x)dx}}. \tag{37}$$

This value for c is finite, real, and $\neq 0$. Hence we obtain

$$(38) \qquad \psi(x) = \frac{\varphi(x)}{\sqrt{\int_a^b \varphi^2(x)dx}}.$$

b) Normalized Orthogonal Systems.—We now propose to determine the constants $c_{\alpha\beta}$ in (35) in such a way that the ψ_α's satisfy the following two conditions:

1) All ψ_α's are normalized:

$$(39) \qquad \int_a^b \psi_\alpha^2(x)dx = 1, \alpha = 1, \ . \ . \ . \ , q$$

2) Two ψ_α's with different subscripts are orthogonal:

$$(40) \quad \int_a^b \psi_\alpha(x)\psi_\beta(x)dx = 0, \ \alpha \neq \beta, \ (\alpha, \beta = 1, \ . \ . \ . \ , q).$$

For this purpose we put $\psi_1(x) = c\varphi_1(x)$, where

$$c = \frac{1}{\sqrt{\int_a^b \varphi_1^2(x)dx}}.$$

Then the condition $\int_a^b \psi_1^2(x)dx = 1$ is satisfied.

Next choose $\psi_2(x)$ as a linear function of φ_1 and φ_2. On account of (38) it can then be expressed linearly in terms of ψ_1 and φ_2:

$$\psi_2(x) = \alpha_1\psi_1(x) + \alpha_2\varphi_2(x).$$

We now determine α_1 and α_2 so that the conditions 1) and 2) are satisfied for ψ_1 and ψ_2.

From (40)

$$(42) \quad (\psi_1 \ \psi_2) = \alpha_1 \int_a^b \psi_1^2 dx + \alpha_2 \int_a^b \psi_1(x)\varphi_2(x)dx = 0.$$

Therefore, on account of (39),

$$\alpha_1 = - \ \alpha_2(\psi_1 \ \varphi_2)$$

and (41) becomes

$$\psi_2(x) = \alpha_2\Big[\varphi_2(x) - (\psi_1\varphi_2)\psi_1(x)\Big].$$

We can now determine α_2 in such a way that $\psi_2(x)$ is normalized provided

$$(43) \qquad \varphi_2(x) - (\psi_1\varphi_2)\psi_1(x) \not\equiv 0.$$

But $(\psi_1\varphi_2)$ is a constant and if (43) is not true, then

$$\varphi_2(x) + c_1\varphi_1(x) \equiv 0$$

which is a linear relation between two fundamental functions φ_1 and φ_2 and, therefore, φ_1 and φ_2 would be linearly dependent, which is contrary to hypothesis. Therefore, $\psi_2(x)$ is completely determined as a linear function of φ_1 and φ_2 or of ψ_1 and φ_2. Now choose $\psi_3(x)$ as a linear function of φ_1, φ_2, φ_3, or, what amounts to the same thing, as a linear function of ψ_1, ψ_2, φ_3:

$$\psi_3(x) = \beta_1\psi_1 + \beta_2\psi_2 + \beta_3\varphi_3.$$

Apply condition (2). We obtain

$$(\psi_1\psi_3) = \beta_1(\psi_1\psi_1) + \beta_2(\psi_1\psi_2) + \beta_3(\psi_1\varphi_3) = 0.$$

But $(\psi_1\psi_1) = 1$ and $(\psi_1\psi_2) = 0$, therefore

$$(\psi_1\psi_3) = \beta_1 + \beta_2(\psi_1\varphi_3) = 0.$$

Hence

$$\beta_1 = -\beta_3(\psi_1\varphi_3).$$

Also by condition (2) we have

$$(\psi_2\psi_3) = \beta_1(\psi_2\psi_1) + \beta_2(\psi_2\psi_2) + \beta_3(\psi_2\psi_3) = 0,$$

from which we derive

$$\beta_2 = -\beta_3(\psi_2\varphi_3).$$

The expression for $\psi_3(x)$ now becomes

$$\psi_3(x) = \beta_3\Big[\varphi_3 - (\psi_1\varphi_3)\psi_1 - (\psi_2\varphi_3)\psi_2\Big].$$

We can now determine β_3 by means of condition 1), provided it is not true that

$$\varphi_3 - (\psi_1\varphi_3)\psi_1 - (\psi_2\varphi_3)\psi_2 \equiv 0. \tag{44}$$

But (44) cannot be satisfied, otherwise φ_1, φ_2, φ_3 would be linearly dependent, which is contrary to hypothesis. Hence β_3 can be so determined that $\psi_3(x)$ satisfies the conditions 1) and 2). This process can be continued until q functions $\psi_\alpha(x)$ are obtained, satisfying conditions 1) and 2), or, as we say, form a *normalized orthogonal system.*

The ψ_α's so obtained are real, they are furthermore linearly independent. For, suppose we had

$$c_1\psi_1(x) + c_2\psi_2(x) + \cdot\cdot\cdot + c_q\psi_q(x) \equiv 0.$$

Multiply by $\psi_\alpha(x)$ and integrate from a to b. On account of (39) and (40) we would obtain $c_\alpha = 0$.

A linear transformation

$$y_i = c_{i1}x_1 + \cdot\cdot\cdot + c_{in}x_n \quad (i = 1, \ldots, n)$$

is said to be orthogonal if

$$\sum_{i=1}^{n} x_i^2 = \sum_{i=1}^{n} y_i^2.$$

The transformation of rectangular coordinates with fixed origin is a transformation of this kind.

The condition on the coefficients which insures this property of the transformation is

$$\sum_{j=1}^{n} c_{ij}c_{kj} = \begin{cases} 1, & i = K \\ 0, & i \neq K. \end{cases} \tag{45}$$

Now consider the definite integral

$$\int_a^b \psi_i(x)\psi_k(x)dx$$

as the limit of a sum

$$(46) \qquad \lim_{h\to 0}\sum_j h\psi_i(x_j)\psi_k(x_j) = \int_a^b \psi_i(x)\psi_k(x)dx.$$

Then the analogy between equations (39), (40), and (45) becomes apparent.

c) Complete Normalized Orthogonal System of Fundamental Functions.—To each root of $D(\lambda) = 0$ there belongs such a normalized orthogonal system of fundamental functions. Now we have shown that $D(\lambda) = 0$ has at least one root. There may be a finite or an infinite number of such roots. If they are infinite in number it follows from the theory of permanently convergent power series that they constitute a *denumerable* set and they may be arranged in the order of the magnitude of their absolute values:

$$|\lambda_1| \leqq |\lambda_2| \leqq \ . \ . \ . \ \leqq |\lambda_n| \leqq |\lambda_{n+1}| \leqq \ . \ . \ .$$

and each λ_i has a definite index q_i. We have then the following table:

Characteristic constant	Index	Normalized orthogonal fundamental functions
λ_1	q_1	$\psi_1^1, \ . \ . \ . \ , \psi_{q_1}^1$
λ_2	q_2	$\psi_1^2, \ . \ . \ . \ , \psi_{q_2}^2$
.	.	. . . .
.	.	. . . .
.	.	. . . .
λ_n	q_n	$\psi_1^n, \ . \ . \ . \ . \ , \psi_{q_n}^n$

We now change the notation and denote the functions ψ in the order in which they stand in the following line:

$$\psi_1^1, \ . \ . \ . \ , \psi_{q_1}^1, \psi_1^2, \ . \ . \ . \ , \psi_{q_2}^2, \ . \ . \ . \ , \psi_1^n, \ . \ . \ . \ , \psi_{q_n}^n$$

by the symbols $\psi_1, \psi_2, \ . \ . \ . \ , \psi_r, \ . \ . \ .$ and the characteristic constants to which they belong by $\lambda_1, \lambda_2, \ . \ . \ . \ ,$

$\lambda_r, \ldots,$ where some of the λ's may be equal, for instance $\lambda_1 = \lambda_2 = \ldots = \lambda_{q_1}; \lambda_{q_1+1} = \lambda_{q_1+2} = \ldots = \lambda_{q_1+q_2}$. Such a system of fundamental functions ψ_r is called a complete normalized orthogonal system of fundamental functions. We sum up the preceding discussion in the following theorem:

Theorem VI.—*To every real symmetric kernel there belongs a complete normalized orthogonal system of fundamental functions $\psi_r(x)$, with the following properties:*

1) *$\psi_r(x)$ is a fundamental function belonging to λ_r*

$$\psi_r(x) = \lambda_r \int_a^b K(x, t)\psi_r(t)dt.$$

2) $\displaystyle\int_a^b \psi_r^2(x)dx = 1.$

3) $\displaystyle\int_a^b \psi_r(x)\psi_s(x)dx = 0, \quad (r \neq s).$

4) *$\psi_r(x)$ is real.*

5) *Every fundamental function $\varphi(x)$ is expressible in the form*

$$\varphi(x) = c_{r_1}\psi_{r_1}(x) + \ldots + c_{r_m}\psi_{r_m}(x).$$

Proof of 3).—If ψ_r, ψ_s belong to the same λ, they are orthogonal by construction, if to different λ's by Theorem III.

Proof of 5).—$\varphi(x)$ belongs to a certain characteristic constant λ_r and is, therefore, linearly expressible in terms of the fundamental functions $\psi_1{}^r(x), \ldots, \psi_q{}^r$, if we use the notation of the table.

Example.—For the problem of the vibrating string we had the kernel:

$$K(x, t) = \begin{cases} (1 - t)x, & 0 \leq x \leq t \\ (1 - x)t, & t \leq x \leq 1 \end{cases} \tag{47}$$

with the characteristic constants $\lambda_n = n^2\pi^2$, $n = 1, 2 \ldots$ of index 1, and the corresponding fundamental functions $\varphi_n(x) = \beta \sin n\pi x$. These functions $\varphi_n(x)$ will form a complete normalized orthogonal system of fundamental functions, if

$$\beta = \frac{1}{\sqrt{\int_0^1 \sin^2 n\pi\, x\, dx}} = \pm\sqrt{2}.$$

Our fundamental functions are then

$$\psi_n(x) = \sqrt{2}\sin n\pi x.$$

III. Expansion of an Arbitrary Function According to the Fundamental Functions of a Complete Normalized Orthogonal System

62. a) Problem of the Vibrating String Resumed.—The problem was to determine $y(x, t)$ so as to satisfy the following conditions (see §26):

$$(48) \qquad 1)\ \frac{\partial^2 y}{\partial t^2} = c^2 \frac{\partial^2 y}{\partial x^2}.$$

$$(49) \qquad 2)\ y(0, t) = 0,\ y(1, t) = 0.$$

$$(50) \qquad 3)\ y(x, 0) = f(x),\ f(x) \text{ an arbitrary given function.}$$
$$y_t(x, 0) = F(x), F(x) \text{ an arbitrary given function.}$$

We attempted to find a solution in the form

$$y = \varphi(t)u(x)$$

and found that $u(x)$ must satisfy

$$1)\ \frac{d^2u}{dx^2} + \lambda u = 0$$

$$2)\ u(0) = 0,\ \ u(1) = 0,$$

and that $\varphi(t)$ must satisfy the equation

$$\frac{d^2\varphi}{dt^2} + \lambda c^2 \varphi = 0.$$

The boundary problem in u had non-trivial solutions only when

$$\lambda = \lambda_n = n^2\pi^2,$$

n a positive integer, the solutions being

$$u = \beta \sin n\pi x.$$

To complete the solution of the partial differential equation (48) with the conditions (49) and (50), we must next integrate the differential equation for $\varphi(t)$ for the values $\lambda = n^2\pi^2$:

$$\frac{d^2\varphi}{dt^2} + n^2\pi^2c^2\varphi = 0.$$

The most general solution of this equation is

$$\varphi(t) = A_n \cos n\pi ct + B_n \sin n\pi ct$$

(A_n, B_n, arbitrary constants), which gives for a solution of the problem of the vibrating string

$$y = (A_n \cos n\pi ct + B_n \sin n\pi ct) \sin n\pi x.$$

This expression for y will satisfy (48) and (49), but, in general, it will not satisfy (50). In order to obtain a solution of (48) which will satisfy both (49) and (50) we notice that, owing to the linear character of (48), the series

$$(51) \qquad y = \sum_{n=1}^{\infty} (A_n \cos n\pi ct + B_n \sin n\pi ct) \sin n\pi x$$

will also satisfy (48) and (49), provided it is convergent and admits of two successive term-by-term differentiations with respect to t and x.

Assuming this condition satisfied, it remains, then, so to determine the constants A_n and B_n that (50) is satisfied:

$$y(x, 0) = \sum_{n=1}^{\infty} A_n \sin n\pi x = f(x)$$

$$y_t(x, 0) = \sum_{n=1}^{\infty} B_n n\pi c \sin n\pi x = F(x).$$

But this is equivalent to asking us to develop the arbitrarily given functions $f(x)$ and $F(x)$ into sine series, or, since

$\psi_n(x) = \sqrt{2}\sin n\pi x$, into a series proceding according to the complete normalized orthogonal system of fundamental functions of the special kernel $K(x, t)$ defined by (47).

b) Determination of the Coefficients in the General Problem. The problem just considered is a particular case of the following more general problem: Given a symmetric kernel $K(x, t)$ with its complete normalized orthogonal system of fundamental functions ψ_ν with corresponding characteristic constants λ_ν, it is possible to expand an arbitrary continuous function $f(x)$ in the form

$$f(x) = \sum_\nu c_\nu \psi_\nu(x).$$

Theorem VII.—*If $f(x)$ is a continuous function and is expressible in the form*

$$f(x) = \sum_\nu c_\nu \psi_\nu(x) \tag{52}$$

and if this series, if infinite, is uniformly convergent on $\left[ab\right]$, then the coefficients c_ν are given by

$$c_n = \int_a^b f(x)\psi_n(x)dx \equiv (f\psi_n)$$

Proof.—Multiply both sides of (52) by $\psi_n(x)$ and integrate with respect to x from a to b:

$$\int_a^b f(x)\psi_n(x) = \int_a^b \sum_\nu c_\nu\psi_\nu(x)\psi_n(x)dx$$
$$= \sum_\nu \int_a^b c_\nu\psi_\nu(x)\psi_n(x)dx$$
$$= c_n,$$

since
$$\int_a^b \psi_\nu(x)\psi_n(x)dx = \begin{cases} 1, & \nu = n \\ 0, & \nu \neq n. \end{cases}$$

IV. Expansion of the Kernel According to the Fundamental Functions of a Complete Normalized Orthogonal System

63. a) Determination of the Coefficients.—$K(x, t)$ with t fixed is a function of x continuous on $\left[ab\right]$. Hence, if we assume that $K(x, t)$ can be expanded into a uniformly convergent series on $\left[ab\right]$:

$$K(x, t) = \sum_{\nu} c_\nu \psi_\nu(x),$$

then, by Theorem VII above, we have

$$c_n = \int_a^b K(x, t)\psi_n(x)dx. \tag{53}$$

But

$$\psi_n(t) = \lambda_n \int_a^b K(t, x)\psi_n(x)dx$$
$$= \lambda_n \int_a^b K(x, t)\psi_n(x)dx,$$

on account of the symmetry of $K(x, t)$. Therefore

$$c_\nu = \frac{\psi_\nu(t)}{\lambda_\nu}$$

and

$$K(x, t) = \sum_{\nu} \frac{\psi_\nu(x)\psi_\nu(t)}{\lambda_\nu}.$$

Thus we obtain the following corollary to Theorem VII.

Corollary.—If $K(x, t) = \sum_{\nu} c_\nu \psi_\nu(x)$ *and this series, if infinite, is uniformly convergent, then*

$$K(x, t) = \sum_{\nu} \frac{\psi_\nu(x)\psi_\nu(t)}{\lambda_\nu}. \tag{54}$$

Equation (54) is what Kneser calls the *bilinear formula.*

If we apply this bilinear formula to

$$K(x, t) = \begin{cases} (1 - t)x, \ 0 \leqq x \leqq t \\ (1 - x)t, \ t \leqq x \leqq 1, \end{cases}$$

we obtain the hypothetical expansion:[1]

$$K(x, t) = \sum_{n=1}^{\infty} \frac{2 \sin n\pi x \sin n\pi t}{n^2\pi^2}.$$

b) The Bilinear Formula for the Case of a Finite Number of Fundamental Functions.—The bilinear formula holds always if the complete normalized orthogonal system contains only a finite number, m, of fundamental functions ψ_ν. The formula to be proved then reads

$$K(x, t) = \sum_{\nu=1}^{m} \frac{\Psi_\nu(x)\Psi_\nu(t)}{\lambda_\nu}. \tag{55}$$

To show that (55) holds, we show that the difference

$$H(x, t) = K(x, t) - \sum_{\nu=1}^{m} \frac{\Psi_\nu(x)\Psi_\nu(t)}{\lambda_\nu} \equiv 0, \text{ on } R. \tag{56}$$

Now $H(x, t)$ is continuous, real, and symmetric. If, in addition $H \not\equiv 0$, then there would exist at least one characteristic constant of $H(x, t)$ considered as a kernel of an integral equation. Hence, if we can show that $H(x, t)$ has no characteristic constant, then it follows that

$$H(x, t) \equiv 0.$$

Suppose that $\varphi(x)$ is a fundamental function of H belonging to a characteristic constant ρ of H, then φ is continuous and $\not\equiv 0$, and

$$\varphi(x) = \rho \int_a^b H(x, t)\varphi(t)dt.$$

[1] This formula is actually true. For a direct proof, see KNESER, "Die Integralgleichungen und ihre Anwendung," §4.

Whence, substituting for $H(x, t)$ its value from (56),

$$(57)\quad \varphi(x) = \rho \int_a^b K(x, t)\varphi(t)dt - \rho \sum_{\nu=1}^{m} \frac{\psi_\nu(x)}{\lambda_\nu} \int_a^b \psi_\nu(t)\varphi(t)dt.$$

Multiply both sides of (57) by $\psi_n(x)$ and integrate with respect to x from a to b, then

$$\begin{aligned}\int_a^b \varphi(x)\psi_n(x)dx &= \rho \int_a^b \int_a^b K(x, t)\psi_n(x)\varphi(t)dtdx \\ &\quad - \rho \sum_{\nu=1}^{m} \frac{1}{\lambda_\nu} \int_a^b \psi_\nu(x)\psi_n(x)dx \int_a^b \psi_\nu(t)\varphi(t)dt \\ &= \frac{\rho}{\lambda_n} \int_a^b \psi_n(t)\varphi(t)dt - \frac{\rho}{\lambda_n} \int_a^b \psi_n(t)\varphi(t)dt \\ &= 0, \qquad (n = 1, \ . \ . \ . \ , m).\end{aligned}$$

Therefore, from (57),

$$\varphi(x) = \rho \int_a^b K(x, t)\varphi(t)dt.$$

By hypothesis $\varphi(x)$ is continuous and $\not\equiv 0$, and so this equation shows that $\varphi(x)$ is a fundamental function of $K(x, t)$ belonging to ρ. Therefore

$$\varphi(x) = c_1\psi_1(x) + \ . \ . \ . \ + c_m\psi_m(x),$$

whence

$$\begin{aligned}\int_a^b \varphi(x)\psi_n(x)dx &= \int_a^b \psi_n(x) \sum_{\nu=1}^{m} c_\nu\psi_\nu(x)dx \\ &= c_n = 0.\end{aligned}$$

Therefore $\varphi(x) \equiv 0$, which constitutes a contradiction. Therefore $H(x, t)$ has no characteristic constant and $H(x, t) \equiv 0$, and hence the bilinear formula holds for m finite. This gives us the following theorem:

Theorem IX.—*If there are a finite number m of characteristic constants λ_ν, then*

$$K(x, t) = \sum_{\nu=1}^{m} \frac{\psi_\nu(x)\psi_\nu(t)}{\lambda_\nu}.$$

c) The Bilinear Formula for Kernels Having an Infinite Number of Characteristic Constants.—We proceed to prove the following theorem:

Theorem X.—*If there are an infinite number of characteristic constants and* $\sum_{\nu=1}^{\infty} \frac{\psi_\nu(x)\psi_\nu(t)}{\psi_\nu}$ *is uniformly convergent in x and t on R, then*

$$K(x, t) = \sum_{\nu=1}^{\infty} \frac{\psi_\nu(x)\psi_\nu(t)}{\lambda_\nu}.$$

Proof.—The proof is entirely analogous to the proof of Theorem IX. Form

$$H(x, t) \equiv K(x, t) - \sum_{\nu=1}^{\infty} \frac{\psi_\nu(x)\psi_\nu(t)}{\lambda_\nu}. \tag{58}$$

H is continuous, for K is continuous, and the sum of the infinite series in (58) is a continuous function, being a uniformly convergent series the terms of which are continuous.[1] Furthermore, H is real, for K and ψ_ν are real. Finally, H is symmetric, for K is symmetric, and each term of the infinite series in (58) is symmetric. Hence it follows, as under *b*), that if $H(x, t)$ has no characteristic constant, then $H \equiv 0$. Suppose $H(x, t)$ had a characteristic constant ρ. Let $\varphi(x)$ be a fundamental function for H, belonging to the characteristic constant ρ, which implies that $\varphi(x)$ is continuous and $\not\equiv 0$ and satisfies

$$\varphi(x) = \rho \int_a^b H(x, t)\varphi(t)dt,$$

which, on account of (58), may be written

$$\varphi(x) = \rho \int_a^b K(x, t)\varphi(t)dt - \rho \int_a^b \sum_{\nu=1}^{\infty} \frac{\psi_\nu(x)\psi_\nu(t)}{\lambda_\nu}\varphi(t)dt. \tag{59}$$

[1] GOURSAT-HEDRICK, "Mathematical Analysis," vol. 1, §173.

The series in (59) is again uniformly convergent as to x in $\left[ab\right]$. Multiply both sides of (59) by $\psi_n(x)$ and integrate with respect to x from a to b. We obtain

$$\int_a^b \psi_n(x)\varphi(x)dx = \rho \int_a^b \int_a^b K(x, t)\varphi(t)\psi_n(x)dtdx$$
$$- \rho \sum_{\nu=1}^{\infty} \frac{1}{\lambda_\nu} \int_a^b \psi_\nu(x)\psi_n(x)dx \int_a^b \psi_\nu(t)\varphi(t)dt,$$

which reduces to

$$\int_a^b \psi_n(x)\varphi(x)dx = \frac{\rho}{\lambda_n}\int_a^b \psi_n(t)\varphi(t)dt - \frac{\rho}{\lambda_n}\int_a^b \psi_n(t)\varphi(t)dt$$
$$= 0, \ (n).$$

Therefore, from (59),

$$\varphi(x) = \rho \int_a^b K(x, t)\varphi(t)dt.$$

Hence we infer, as under b), that $\varphi(x) \equiv 0$, which contradicts our assumption $\varphi(x) \not\equiv 0$. Therefore, $H(x, t)$ has no characteristic constant and, hence, $H(x, t) \equiv 0$.

Since a series with a finite number of terms, functions of x and t, is always uniformly convergent, we may combine Theorems IX and X into the following

Theorem XI.—*If* $\left\{ \psi_\nu(x) \right\}$ *are a complete normalized orthogonal system of fundamental functions belonging to the characteristic constant* λ_ν, *for the real symmetric kernel* $K(x, t)$, *and*

$$\frac{1}{\lambda_\nu}\sum_\nu \psi_\nu(x)\psi_\nu(t)$$

is uniformly convergent on R, *then*

$$K(x, t) = \sum_\nu \frac{\psi_\nu(x)\psi_\nu(t)}{\lambda_\nu}.$$

64. The Complete Normalized Orthogonal System for the Iterated Kernel $K_n(x, t)$.—The series $\sum_\nu \frac{\psi_\nu(x)\psi_\nu(t)}{\lambda_\nu}$ is not always uniformly convergent and therefore the bilinear formula does not hold for every kernel $K(x, t)$. But we shall be able to show that it does always hold for the iterated kernels. For this purpose we prove first the following:

Theorem XII.—*If* $\left\{ \psi_\nu(x) \right\}$ *are a complete normalized orthogonal system of fundamental functions for* $K(x, t)$ *belonging to the characteristic constants* $\left\{ \lambda_\nu \right\}$, *then* $\left\{ \psi_\nu(x) \right\}$ *are a complete normalized orthogonal system of fundamental functions for* $K_n(x, t)$ *belonging to the characteristic constants* $\left\{ \lambda_\nu^n \right\}$.

Proof.—*a*) By assumption, $\psi_\nu(x)$ is continuous and $\not\equiv 0$ and

$$\psi_\nu(x) = \lambda_\nu \int_a^b K(x, t)\psi_\nu(t)dt, \tag{61}$$

whence we derive

$$\int_a^b K(z, x)\psi_\nu(x)dx = \lambda_\nu \int_a^b \int_a^b K(z, x)K(x, t)\psi_\nu(t)dtdx,$$

which, from (61) and our definitions of iterated kernels, may be written

$$\frac{\psi_\nu(z)}{\lambda_\nu} = \lambda_\nu \int_a^b K_2(z, t)\psi_\nu(t)dt,$$

or, if we multiply by λ_ν and put x in place of z,

$$\psi_\nu(x) = \lambda_\nu^2 \int_a^b K_2(x, t)\psi_\nu(t)dt. \tag{62}$$

From (62) we derive

$$\psi_\nu(x) = \lambda_\nu^3 \int_a^b K_3(x, t)\psi_\nu(t)dt$$

in the same way that (62) was derived from (61). Continuing this process, it is easy to show by mathematical induction that

$$\psi_\nu(x) = \lambda_\nu{}^n \int_a^b K_n(x, t)\psi_\nu(t)dt. \tag{63}$$

Therefore, $\lambda_\nu{}^n$ is a characteristic constant of $K_n(x, t)$, and $\psi_\nu(x)$ is a fundamental function of $K_n(x, t)$ belonging to $\lambda_\nu{}^n$.

b) It is still necessary to show that $K_n(x, t)$ has no other characteristic constants than the $\lambda_\nu{}^n$, and that every fundamental function $\varphi(x)$ of $K_n(x, t)$ can be expressed in the form

$$\varphi(x) = C_{\nu_1}\psi_{\nu_1}(x) + \ . \ . \ . + C_{\nu_r}\psi_{\nu_r}(x).$$

Let ρ be any characteristic constant of $K_n(x, t)$ and $\varphi(x)$ be a fundamental function of $K_n(x, t)$ belonging to ρ:

$$\varphi(x) = \rho \int_a^b K_n(x, t)\varphi(t)dt, \ \varphi(x)^{c.} \not\equiv 0. \tag{64}$$

We must then prove that

1) $\rho = \lambda_\nu{}^n$.

2) $\varphi(x) = \sum_{i=1}^{r} C_{\nu_i}\psi_{\nu_i}(x)$.

Let $h_1, h_2, \ . \ . \ . \ , h_n$ be the nth roots of ρ:

$$h_\nu{}^n = \rho, \ (\nu = 1, \ . \ . \ . \ , n). \tag{65}$$

Build up the functions $nG_i(x)$ as follows:

$$\begin{aligned} nG_i(x) = \varphi(x) &+ h_i \int_a^b K(\imath, t)\varphi(t)dt \\ &+ h_i^2 \int_a^b K_2(x, t)\varphi(t)dt + \ . \ . \ . \\ &+ h_i^{n-1} \int_a^b K_{n-1}(x, t)\varphi(t)dt, \end{aligned} \tag{66}$$

for $i = 1, \ . \ . \ . \ , n$. Add these equations and cancel a factor n. We obtain

$$G_1(x) + \ . \ . \ . + G_n(x) = \varphi(x), \tag{67}$$

since $h_1^s + h_2^s + \ldots + h_n^s = 0,$

$$(s = 1, \ldots, n - 1).$$

Multiply both sides of (66) by $K(z, x)$ and integrate. We obtain

$$n\int_a^b K(z, x)G_i(x)dx = \int_a^b K(z, x)\varphi(x)dx + \sum_{s=1}^{n-1} h_i^s \int_a^b\int_a^b K(z, x)K_s(x, t)\varphi(t)dtdx.$$

Multiply the members of this equation by h_i, put $h_i^n = \rho$ by (65), make use of (64) for $x = z$, $t = x$, and of the definitions of the iterated kernels. We obtain

$$nh_i\int_a^b K(z, x)G_i(x)dx = h_i\int_a^b K(z, t)\varphi(t)dt + h_i^2\int_a^b K_2(z, t)\varphi(t)dt + \ldots + h_i^{n-1}\int_a^b K_{n-1}(z, t)\varphi(t)dt + \varphi(z),$$

which, on account of (66), may be written, after canceling a factor n,

$$h_i\int_a^b K(z, x)G_i(x)dx = G_i(z).$$

Suppose for some value of i that $G_i(x) \not\equiv 0$, then h_i is a characteristic constant of $K(x, t)$ and $G_i(x)$ is a fundamental function of $K(x, t)$ belonging to h_i. Therefore, there exists a value m of ν, such that $h_i = \lambda_m$ and hence $\rho = \lambda_m^n$. Since $G_i(x)$ is a fundamental function of $K(x, t)$, it can be written in the form

$$(68) \qquad G_i(x) = C_{i_1}\psi_{i_1} + \ldots + C_{i_k}\psi_{i_k}.$$

But the $G_i(x)$ do not all vanish identically, otherwise from (67) we would have $\varphi(x) \equiv 0$, which cannot be. Hence

$$\rho = \lambda_m^n,$$

and from (67) and (68) we have

$$\varphi(x) = C_{\nu_1}\psi_{\nu_1}(x) + \cdot \cdot \cdot + C_{\nu_r}\psi_{\nu_r},$$

which completes the proof of the theorem.

V. Auxiliary Theorems

65. Bessel's Inequality.—For the further discussion of the bilinear formula we need the following theorem:

Theorem XIII.—*Given*

1) $f(x)$, *real and continuous.*

2) ψ_s, $(s = 1, \ldots, m)$, *real and continuous, constituting a normalized orthogonal set:*

$$\int_a^b \psi_r\psi_s dx = \begin{cases} 1, & r = s \\ 0, & r \neq s, \end{cases}$$

then

$$(69) \qquad \sum_{r=1}^{m}\left[\int_a^b f(x)\psi_s(x)dx\right]^2 \leq \int_a^b \left[f(x)\right]^2 dx.$$

Proof.—Let C_s be *any* real constants, then

$$\int_a^b \left[f(x) - \sum_{s=1}^{m} C_s\psi_s(x)\right]^2 dx \geq 0,$$

whence, after squaring the expression in the bracket,

$$(70) \quad \int_a^b \left[f(x)\right]^2 dx - 2\sum_{s=1}^{m} C_s \int_a^b f(x)\psi_s(x)dx + \int_a^b \left[\sum_{s=1}^{m} C_s\psi_s(x)\right]^2 dx \geqq 0.$$

But

$$\int_a^b \left[\sum_{s=1}^{m} C_s\psi_s(x)\right]^2 dx = \sum_{s=1}^{m} C_s^2 \int_a^b \psi_s^2 dx + 2\sum_{(r,s)} C_r C_s \int_a^b \psi_r\psi_s dx.$$

Now

$$\int_a^b \psi_s^2 dx = 1 \text{ and } \int_a^b \psi_r \psi_s dx = 0$$

by hypothesis. Therefore,

$$\int_a^b \left[\sum_{s=1}^m C_s\psi_s(x)\right]^2 dx = \sum_{s=1}^m C_s^2.$$

But C_s are any real constants and so we may choose

$$C_s = \int_a^b f(x)\psi_s(x)dx.$$

For this choice of C_s, equation (70) becomes

$$\int_a^b \left[f(x)\right]^2 dx - 2\sum_{s=1}^m \left(\int_a^b f(x)\psi_s(x)dx\right)^2 + \sum_{s=1}^m \left(\int_a^b f(x)\psi_s(x)dx\right)^2 \geqq 0,$$

whence we obtain (69).

We now apply (69) to the particular case where

$$f(x) = K(x, t)$$

for a fixed t, and the functions ψ_s are a normalized orthogonal system of fundamental functions of $K(x, t)$ belonging to λ_s.

Equation (69) now becomes

$$(71) \qquad \sum_{s=1}^m \left(\int_a^b K(x, t)\psi_s(x)dx\right)^2 \leqq \int_a^b \left[K(x, t)\right]^2 dx.$$

But

$$\psi_s(x) = \lambda_s \int_a^b K(x, t)\psi_s(t)dt,$$

whence

$$\psi_s(t) = \lambda_s \int_a^b K(t, x)\psi_s(x)dx$$

$$(72) \qquad = \lambda_s \int_a^b K(x, t)\psi_s(x)dx,$$

on account of the symmetry of $K(x, t)$. Now $K(x, t)$ is continuous on R and, therefore,

(73) $$|K(x, t)| \leqq G \quad \text{(a constant)}.$$

Applying (72) and (73) to (71), we obtain

$$\sum_{s=1}^{m}\left(\frac{\psi_s(t)}{\lambda_s}\right)^2 \leqq G^2(b - a).$$

Thus we have the following corollary to Theorem XIII:

Corollary.—If $\psi_s(x)$, $(s = 1, \ldots, m)$ *are normalized orthogonal fundamental functions of the kernel* $K(x, t)$, *belonging to the characteristic constant* λ_s, *then*

(74) $$\sum_{s=1}^{m}\left(\frac{\psi_s(x)}{\lambda_s}\right)^2 \leqq G^2(b - a)$$

From (74) we obtain by integrating from a to b

$$\sum_{s=1}^{m}\frac{1}{\lambda^2_s}\int_a^b\Big[\psi_s(x)\Big]^2 dx \leqq G^2(b - a)^2.$$

But $$\int_a^b\Big[\psi_s(x)\Big]^2 dx = 1 \text{ by hypothesis.}$$

Therefore

(75) $$\sum_{s=1}^{m}\frac{1}{\lambda_s^2} \leqq G^2(b - a)^2,$$

the inequality holding for *any* finite number of λ_s and hence for the first m of the λ_s's.

Theorem XIV.—$\sum_{s=1}\frac{1}{\lambda_s^2}$ *is convergent.*

Proof.—The proof follows from (75) by applying the principle of monotony: if u_i are real and positive and

$$u_1 + u_2 + \ldots + u_n \leqq A, \ (n)$$

that is bounded, then

$$\sum_{n=1}^{\infty} u_n$$

is convergent.

66. Proof of the Bilinear Formula for the Iterated Kernel $K_n(x, t)$ for $n \leq 4$.—In Theorem XI it was stated that if $\sum_\nu \frac{\psi_\nu(x)\psi_\nu(t)}{\lambda_\nu}$ was uniformly convergent, then

$$\sum_\nu \frac{\psi_\nu(x)\psi_\nu(t)}{\lambda_\nu} = K(x, t).$$

We have stated previously that this series is not always uniformly convergent and, therefore, the bilinear formula does not hold for every kernel. But we shall be able to prove that it does always hold for the iterated kernels.

Since, according to Theorem XII, $\psi_\nu(x)$ are a complete normalized orthogonal system of fundamental functions for $K_n(x, t)$ belonging to $\lambda_\nu{}^n$, the bilinear formula for the iterated kernel, if true, would read

$$\sum_\nu \frac{\psi_\nu(x)\psi_\nu(t)}{\lambda_\nu{}^n} = K_n(x, t), \ n > 1. \tag{76}$$

By referring to Theorem XI just quoted, we see that (76) will be proved if we can show that $\sum_\nu \frac{\psi_\nu(x)\psi_\nu(t)}{\lambda_\nu{}^n}$ is uniformly convergent. We prove this first for the case $n = 4$. We desire then to show the uniform convergence of

$$\sum_\nu \frac{\psi_\nu(x)\psi_\nu(t)}{\lambda_\nu{}^4} \text{ on } R. \tag{77}$$

We apply the general test for the uniform convergence of the series

$$\sum_{\nu=1}^{\infty} u_\nu(x, t), \text{ on } R.$$

For any $\epsilon > 0$ we can assign an N depending on ϵ, but not on x and t (N_ϵ) such that

$$|u_{n+1}(x, t) + u_{n+2}(x, t) + \ . \ . \ . + u_{n+p}(x, t)| < \epsilon, \ (R)$$

for every $n > N_\epsilon$ for any integer $p > 0$.

Use the notation

$$\Delta_{np} = \sum_{\nu=n+1}^{n+p} \frac{\psi_\nu(x)\psi_\nu(t)}{\lambda_\nu^{\,4}}.$$

We have the inequalities

$$(78)\quad |\Delta_{np}| \leq \sum_{\nu=n+1}^{n+p} \frac{|\psi_\nu(x)\psi_\nu(t)|}{\lambda_\nu^{\,4}} \leq \sum_\nu \frac{|\psi_\nu(x)||\psi_\nu(t)|}{|\lambda_\nu||\lambda_\nu|} \sum_\nu \frac{1}{\lambda_\nu^{\,2}}.$$

We now make use of the inequality

$$(79)\qquad \sum a_\nu b_\nu \leq \sqrt{\sum a_\nu^2}\sqrt{\sum b_\nu^2}$$

(a_ν, b_ν real and positive), which arises from the following consideration: given the matrix

$$\left\| \begin{matrix} a_1\, a_2 & \ldots & a_p \\ b_1\, b_2 & \ldots & b_p \end{matrix} \right\|, \text{ form the product}$$

$$(80)\quad \left\| \begin{matrix} a_1\, a_2 & \ldots & a_p \\ b_1\, b_2 & \ldots & b_p \end{matrix} \right\|^2 = \begin{vmatrix} \sum_{\nu=1}^{p} a_\nu^2 & \sum_{\nu=1}^{p} a_\nu b_\nu \\ \sum_{\nu=1}^{p} a_\nu b_\nu & \sum_{\nu=1}^{p} b_\nu^2 \end{vmatrix}$$

$$= \sum_{(\mu,\nu)=1}^{p} \begin{vmatrix} a_\mu & a_\nu \\ b_\mu & b_\nu \end{vmatrix}^2 \geq 0.$$

From the inequality in (80) follows (79), which is the algebraic analogue of Schwarz's inequality (21) of §57. Now apply (79) to (78) with $a_\nu \equiv \left|\frac{\psi_\nu(x)}{\lambda_\nu}\right|$, $b_\nu \equiv \left|\frac{\psi_\nu(t)}{\lambda_\nu}\right|$. We obtain

$$\sum_{\nu=n+1}^{n+p} \left|\frac{\psi_\nu(x)}{\lambda_\nu}\right| \cdot \left|\frac{\psi_\nu(t)}{\lambda_\nu}\right| \leq \sqrt{\sum_{\nu=n+1}^{n+p} \frac{\psi_\nu^{\,2}(x)}{\lambda_\nu^{\,2}}} \sqrt{\sum_{\nu=n+1}^{n+p} \frac{\psi_\nu^{\,2}(t)}{\lambda_\nu^{\,2}}}.$$

But by (74)

$$(81)\quad \sum_{\nu=n+1}^{n+p} \frac{\psi^2(x)}{\lambda_\nu^{\,2}} \leq G^2(b-a), \sum_{\nu=n+1}^{n+p} \frac{\psi_\nu^{\,2}(t)}{\lambda_\nu^{\,2}} \leq G^2(b-a).$$

By Theorem XIV, $\sum \frac{1}{\lambda_\nu^2}$ is convergent, and thus, by the general theorem on convergent series,

$$\lim_{n \to \infty} d_{np} = 0,$$

where

$$d_{np} = \sum_{\nu = n+1}^{n+p} \frac{1}{\lambda_\nu^2}. \tag{82}$$

Substitution from (81) and (82) in (78) gives

$$| \Delta_{np} | \leq G^2(b - a) d_{np}.$$

The right member of this inequality is independent of x and t, therefore (77) is absolutely and uniformly convergent in x and t on R. Thus we have proved the following theorem:

Theorem XV.—$\sum_\nu \frac{\psi_\nu(x)\psi_\nu(t)}{\lambda_\nu^4}$ *is absolutely and uniformly convergent in x and t on R, and*

$$\sum_\nu \frac{\psi_\nu(x)\psi_\nu(t)}{\lambda_\nu^4} = K_4(x, t).$$

We can now show that (76) holds for $n > 4$. Suppose that (76) holds in the nth instance:

$$K_n(x, t) = \sum_\nu \frac{\psi_\nu(x)\psi_\nu(t)}{\lambda_\nu^n}, \tag{83}$$

the series being uniformly convergent in x and t on R. Multiply both sides of (83) by $K(z, x)$ and integrate with respect to x from a to b. We obtain

$$\int_a^b K(z, x) K_n(x, t) dx = \sum_\nu \frac{\psi_\nu(t)}{\lambda_\nu^n} \int_a^b K(z, x) \psi_\nu(x) dx \tag{84}$$

and the series on the right is again uniformly convergent as to x and t. But

$$\int_a^b K(z, x) \psi_\nu(x) dx = \frac{\psi_\nu(z)}{\lambda_\nu}$$

and

$$\int_a^b K(z, x)K_n(x, t)dx = K_{n+1}(z, t).$$

We obtain, then, from (84), if we put $z = x$,

$$K_{n+1}(x, t) = \sum_\nu \frac{\psi_\nu(x)\psi_\nu(t)}{\lambda_\nu^{n+1}}.$$

The induction is now complete and (76) holds for $n \geq 4$. Kowalewski[1] gives a proof that (76) holds for $n = 2$, whence from our induction proof, (76) holds for $n = 3$, and therefore for $n > 1$.

67. An Auxiliary Theorem of Schmidt.—Before we can take up the problem of the expansion of an arbitrary function according to the fundamental functions of a symmetric kernel, we need another auxiliary theorem due to E. Schmidt.

Theorem XVI.—*Let* $h(x)$ *be continuous on* $[ab]$, $K(x, t)$ *real and symmetric with the complete normalized orthogonal system of fundamental functions* $\left\{\psi_\nu(x)\right\}$ *belonging to the characteristic constants* $\left\{\lambda_\nu\right\}$, *then*

A) If $\int_a^b K(x, t)h(t)dt = 0$, *uniformly as to* x, *then*

$$\int_a^b \psi_n(x)h(x)dx = 0, \text{ uniformly as to } n.$$

B) If $\int_a^b \psi_n(x)h(x)dx = 0$, *uniformly as to* n, *then*

$$\int_a^b K(x, t)h(t)dt = 0, \text{ uniformly as to } x.$$

Proof.—*A*) Multiply both members of

$$\int_a^b K(x, t)h(t)dt = 0, \ (x)$$

[1] Kowalewski, "Einführung in die Déterminanten Theorie," Teubner, p. 533, §200, 1900.

by $\psi_n(x)$ and integrate. We obtain

$$\int_a^b \int_a^b K(x, t)\psi_n(x)h(t)dtdx = 0, \tag{85}$$

which may be written

$$\int_a^b h(t)\left(\int_a^b K(x, t)\psi_n(x)dx\right)dt = 0.$$

But
$$\int_a^b K(x, t)\psi_n(x)dx = \frac{\psi_n(t)}{\lambda_n}.$$

Therefore (85) may be written

$$\frac{1}{\lambda_n}\int_a^b \psi_n(t)h(t)dt = 0, \ (n).$$

Therefore
$$\int_a^b \psi_n(t)h(t)dt = 0, \ (n).$$

B) Our hypothesis is that

$$\int_a^b \psi_n(x)h(x)dx = 0, \ (n).$$

We have previously shown that

$$K_4(x, t) = \sum_\nu \frac{\psi_\nu(x)\psi_\nu(t)}{\lambda_\nu^{\,4}}$$

is uniformly convergent on R. Multiply both members of this equality by $h(x)h(t)$ and integrate. We obtain

$$\int_a^b \int_a^b K_4(x, t)h(x)h(t)dxdt = \sum_\nu \frac{1}{\lambda_\nu^{\,4}} \int_a^b \int_a^b \psi_\nu(x) \times$$
$$h(x)\cdot\psi_\nu(t)h(t)dtdx.$$

The right member of this equality may be written

$$\sum_\nu \frac{1}{\lambda_\nu^{\,4}} \int_a^b \psi_\nu(x)h(x)dx \cdot \int_a^b \psi_\nu(t)h(t)dt,$$

which, on account of our hypothesis, vanishes. Therefore

$$\int_a^b \int_a^b K_4(x, t)h(x)h(t)dxdt = 0. \tag{86}$$

But
$$K_4(x, t) = \int_a^b K_2(x, s)K_2(t, s)ds,$$

since $K_2(t, s) = K_2(s, t)$ from the symmetry of K. Substitute this value of $K_4(x, t)$ in (86). We obtain

$$\int_a^b \int_a^b \int_a^b K_2(x, s)K_2(t, s)h(x)h(t)dsdxdt = 0,$$

which may be written

$$\int_a^b \left[\int_a^b K_2(x, s)h(x)dx \times \int_a^b K_2(t, s)h(t)dt \right] ds = 0,$$

or

$$\int_a^b \left[\int_a^b K_2(x, s)h(x)dx \right]^2 ds = 0.$$

But $\int_a^b K_2(x, s)h(x)ds$ is a real continuous function of s. Therefore

$$\int_a^b K_2(x, s)h(x)dx \equiv 0, \ (s),$$

which, if we change our notation and, in place of s and x, put x and t, and remember that $K_2(x, t) = K_2(t, x)$, becomes

$$\int_a^b K_2(x, t)h(t)dt \equiv 0, \ (x).$$

Multiply both members of this last equation by $h(x)$ and integrate. We obtain

$$\int_a^b \int_a^b K_2(x, t)h(x)h(t)dxdt = 0. \tag{87}$$

But

$$K_2(x, t) = \int_a^b K(x, s)K(t, s)ds.$$

Treat (87) in the same manner as we have just done with (86) and we will finally obtain

$$\int_a^b K(x, t)h(t)dt \equiv 0, \ (x).$$

This completes the proof of Schmidt's auxiliary theorem.

VI. Expansion of an Arbitrary Function According to the Complete Normalized Orthogonal System of Fundamental Functions of a Symmetric Kernel

68. We are now in a position to solve the fundamental problem stated in §62: Given an arbitrary function $f(x)$, to represent it, if possible, in the form

$$f(x) = \sum_{\nu} C_\nu \psi_\nu(x),$$

when the $\psi_\nu(x)$ are the fundamental functions of a complete normalized orthogonal system of the symmetric kernel $K(x, t)$.

We have seen in §62, that if such an expansion is possible and, moreover, uniformly convergent, then the coefficients C_ν must have the values

$$C_\nu = \int_a^b f(x)\psi_\nu(x)dx \equiv (f\psi_\nu),$$

that is
$$f(x) = \sum_{\nu} (f\psi_\nu)\psi_\nu(x).$$

We now prove the following theorem:

Theorem XVII.—*If $f(x)$ can be represented in the form*

$$f(x) = \int_a^b K(x, t)g(t)dt, \tag{88}$$

where $g(x)$ is a function continuous on $\left[ab\right]$, then

$$f(x) = \sum_{\nu} (f\psi_\nu)\psi_\nu(x) \text{ on } \left[ab\right]$$

and the series is uniformly and absolutely convergent on $\left[ab\right]$.

Proof.—*a) Convergence Proof.* Use the notation

$$\Delta_{np} = \sum_{\nu=n+1}^{n+p} (f\psi_\nu)\psi_\nu(x), \; p > 0.$$

Then, from the general convergence principle stated in §66, the series

$$\sum_{\nu=1}^{\infty}(f\psi_\nu)\psi_\nu(x) \tag{89}$$

converges uniformly on $\left[ab\right]$ if for every $\epsilon > 0$ there exists an N dependent upon ϵ, but not on $x(N_\epsilon)$, such that for every $n > N_\epsilon$ it is true that $|\Delta_{np}| < \epsilon$ on $\left[ab\right]$. Now

$$\begin{aligned}(f\psi_\nu) &= \int_a^b f(x)\psi_\nu(x)dx \\ &= \int_a^b\int_a^b K(x, t)g(t)\psi_\nu(x)dt\,dx \qquad (90)\end{aligned}$$

on account of (88).

The right member of (90) may be written

$$\int_a^b g(t)\left(\int_a^b K(x, t)\psi_\nu(x)dx\right)dt.$$

But
$$\int_a^b K(x, t)\psi_\nu(x)dx = \frac{\psi_\nu(t)}{\lambda_\nu}.$$

Therefore (90) becomes

$$(f\psi_\nu) = \frac{1}{\lambda_\nu}(g\psi_\nu). \tag{91}$$

Now

$$|\Delta_{np}| \leq \sum_{\nu=n+1}^{n+p} |(f\psi_\nu)|\,|\psi_\nu(x)|.$$

Therefore, by (91),

$$|\Delta_{np}| \leq \sum_{\nu=n+1}^{n+p} |(g\psi_\nu)|\cdot\left|\frac{\psi_\nu(x)}{\lambda_\nu}\right|. \tag{92}$$

But from the algebraic analogue (79) of Schwarz's inequality (21), proved in §57, we have

$$\sum_\nu |(g\psi_\nu)|\left|\frac{\psi_\nu(x)}{\lambda_\nu}\right| \leq \sqrt{\sum_\nu (g\psi_\nu)^2}\sqrt{\sum_\nu \frac{\psi_\nu^{\,2}(x)}{\lambda_\nu^{\,2}}},$$

and from (74) in the corollary to Bessel's inequality

$$\sum_{\nu} \frac{\psi_\nu{}^2(x)}{\lambda_\nu{}^2} \leqq G^2(b-a),$$

G being the maximum of $|K(x, t)|$ on R.

If, now, we use the notation

$$d_{np} = \sum_{\nu=n+1}^{n+p} (g\psi_\nu)^2$$

we may write

(93) $$|\Delta_{np}| \leqq G\sqrt{(b-a)d_{np}}.$$

But

(94) $$\sum_{\nu=1}^{\infty} (g\psi_\nu)^2$$

is convergent, for

$$\sum_{\nu=1}^{n} (g\psi_\nu)^2 = \sum_{\nu=1}^{n} \left(\int_a^b g(x)\psi_\nu(x)dx \right)^2 \leqq \int_a^b \Big[g(x) \Big]^2 dx,$$

from Bessel's inequality (69), §65, and $\int_a^b \Big[g(x) \Big]^2 dx$ is a finite fixed constant, independent of n, and therefore

$$\sum_{\nu=1}^{n} (g\psi_\nu)^2$$

is bounded for every n, and hence, by the monotony principle (94), is convergent. Therefore, $\lim_{n \to \infty} d_{np} = 0$. Hence, from (93), $\lim_{n \to \infty} |\Delta_{np}| = 0$, and therefore (89) is uniformly convergent on $\Big[ab \Big]$, and from (92) we see that it is also absolutely convergent. This completes the first part of the proof. It remains to determine the sum.

b) Summation.—Put

$$h(x) \equiv f(x) - \sum_{\nu} (f\psi_\nu)\psi_\nu(x).$$

From (88), $f(x)$ is continuous, since $g(x)$ is supposed to be continuous and the ψ_ν's are continuous and the series is uniformly convergent, hence h is continuous. We will have proved that the series represents $f(x)$, if we can show

that $$\int_a^b \left[h(x)\right]^2 dx = 0,$$
for then
$$h(x) \equiv 0.$$
Now
$$\int_a^b \left[h(x)\right]^2 dx = (fh) - \sum_\nu (f\psi_\nu)(h\psi_\nu). \tag{95}$$

We compute
$$(h\psi_n) = (f\psi_n) - \sum_\nu (f\psi_\nu)(\psi_\nu\psi_n).$$
But
$$(\psi_\nu\psi_n) = \begin{cases} 1, & \nu = n \\ 0, & \nu \neq n. \end{cases}$$
Therefore
$$(h\psi_n) = (f\psi_n) - (f\psi_n) = 0. \tag{96}$$
Whence, from (95),
$$\int_a^b \left[h(x)\right]^2 dx = (fh). \tag{97}$$
But
$$(fh) = \int_a^b f(x)h(x)dx$$
and, by hypothesis, as given by (88),
$$f(x) = \int_a^b K(x, t)g(t)dt.$$
Therefore,
$$(fh) = \int_a^b \int_a^b K(x, t)g(t)h(x)dtdx$$
$$= \int_a^b g(t)\left(\int_a^b K(x, t)h(x)dx\right)dt. \tag{98}$$
Now
$$\int_a^b \psi_n(x)h(x)dx = 0, \ (n) \text{ by (96)},$$

hence, by Theorem XVI, §67,

$$\int_a^b K(x, t)h(t)dt = 0, \ (x)$$

and, therefore,

$$\int_a^b K(x, t)h(x)dx = 0, \ (t)$$

whence, from (98),

$$(fh) \equiv 0, \ (x)$$

and, therefore, from (97),

$$h(x) \equiv 0, \ (x),$$

which completes the proof of the theorem. We have the following corollaries to Theorem XVII.

Corollary I.—$(f\psi_\nu) = \dfrac{(g\psi_\nu)}{\lambda_\nu}$ by (91).

Corollary II.—If $g(x)$ is continuous, then

$$\int_a^b K(x, t)g(t)dt = \sum_{\nu=1}^{\infty} \frac{(g\psi_\nu)}{\lambda_\nu} \psi_\nu(x) \tag{99}$$

and the series is uniformly and absolutely convergent. This follows from Theorem XVII and Corollary I.

Corollary III.—If $g(x)$ and $h(x)$ are continuous, then

$$\int_a^b \int_a^b K(x, t)g(t)h(x)dtdx = \sum_{\nu=1}^{\infty} \frac{(g\psi_\nu)\,(h\psi_\nu)}{\lambda_\nu}.$$

This is sometimes called *Hilbert's* formula. It follows from (99), if we multiply by $h(x)$ and integrate.

VII. Solution of the Integral Equation

69. Schmidt's Solution of the Non-homogeneous Integral Equation When λ Is Not a Characteristic Constant.—As an application of the last theorem, we will now give Schmidt's solution of the equation

$$f(x) = u(x) - \lambda \int_a^b K(x, t)u(t)dt. \tag{100}$$

A) Necessary Form of the Solution.—Let $u(x)$ be a continuous solution of (100), then

$$\text{(101)} \qquad \mathrm{v}(x) \equiv \frac{u(x) - f(x)}{\lambda} = \int_a^b K(x, t)u(t)dt.$$

The function $\mathrm{v}(x)$ satisfies the condition (88) of Theorem XVII and, therefore,

$$\mathrm{v}(x) = \sum_{\nu} (\mathrm{v}\psi_\nu)\psi_\nu(x),$$

the series being absolutely and uniformly convergent on $\left[ab\right]$. On account of (101) and Corollary I §68, we have

$$\text{(102)} \qquad \mathrm{v}(x) = \sum_{\nu} \frac{(u\psi_\nu)}{\lambda_\nu} \psi_\nu(x).$$

Multiply (100) by $\psi_\nu(x)$ and integrate. We obtain

$$\begin{aligned} (f\psi_\nu) &= (u\psi_\nu) - \lambda \int_a^b \int_a^b K(x, t)u(t)\psi_\nu(x)dtdx \\ &= (u\psi_\nu) - \lambda \int_a^b \frac{\psi_\nu(t)u(t)}{\lambda_\nu} dt \\ &= (u\psi_\nu) - \frac{\lambda}{\lambda_\nu}(u\psi_\nu), \end{aligned}$$

which may be written

$$(\lambda_\nu - \lambda)(u\psi_\nu) = \lambda_\nu(f\psi_\nu).$$

When $\lambda \neq \lambda_\nu$, we have

$$\text{(103)} \qquad (u\psi_\nu) = \frac{\lambda_\nu(f\psi_\nu)}{\lambda_\nu - \lambda}.$$

If λ is not a characteristic constant, then $\lambda \neq \lambda_1, \lambda_2, \ldots$, and for such values of λ (103) holds for all values of ν and hence, from (102),

$$\mathrm{v}(x) = \sum_{\nu=1}^{\infty} \frac{(f\psi_\nu)}{\lambda_\nu - \lambda} \psi_\nu(x).$$

Whence from (100) and (101)

$$u(x) = f(x) + \lambda \sum_{\nu=1}^{\infty} \frac{(f\psi_\nu)}{\lambda_\nu - \lambda} \psi_\nu(x). \tag{104}$$

Hence, if (100) has a continuous solution $u(x)$, this solution is unique and is given by (104). Replacing the symbol $(f\psi_\nu)$ by its explicit expression, we have for the solution of (100)

$$u(x) = f(x) + \lambda \sum_{\nu=1}^{\infty} \left[\frac{1}{\lambda_\nu - \lambda} \int_a^b f(t)\psi_\nu(t)dt \right] \psi_\nu(x). \tag{105}$$

If for a given value of λ, the series

$$\sum_{\nu=1}^{\infty} \frac{\psi_\nu(t)\psi_\nu(x)}{\lambda_\nu - \lambda}$$

is uniformly convergent in t, we can write the solution in a form previously obtained:

$$u(x) = f(x) + \lambda \int_a^b K(x, t; \lambda) f(t) dt$$

where

$$K(x, t; \lambda) = \sum_{\nu=1}^{\infty} \frac{\psi_\nu(x)\psi_\nu(t)}{\lambda_\nu - \lambda}.$$

The solution given by (105) has an advantage over that given by Fredholm, in that it shows the meromorphic character of the solution with respect to λ and indicates the principal part of $u(x)$ with respect to each pole λ.

B) Sufficiency Proof.—It remains to show that (104) is absolutely and uniformly convergent and satisfies (100).

1) *Uniform and Absolute Convergence.*—The series in (104) may be written as follows

$$\sum_{\nu=1}^{\infty} \frac{1}{1 - \dfrac{\lambda}{\lambda_\nu}} \frac{(f\psi_\nu)}{\lambda_\nu} \psi_\nu(x). \tag{106}$$

But by Corollary II to Theorem XVII

$$\sum_\nu \frac{(f\psi_\nu)}{\lambda_\nu}\psi_\nu(x)$$

is absolutely and uniformly convergent; and if there is an infinitude of characteristic constants λ_ν, then as $\nu \to \infty$ also $\lambda_\nu \to \infty$, and therefore $\dfrac{1}{1-\dfrac{\lambda}{\lambda_\nu}} \to 1$, whence also (106) is absolutely and uniformly convergent.

2) (104) *Satisfies* (100).—From (104) we obtain

$$u(x) - \lambda \int_a^b K(x,t)u(t)dt = f(x) + \lambda \sum_\nu \frac{(f\psi_\nu)}{\lambda_\nu - \lambda}\psi_\nu(x)$$
$$- \lambda \int_a^b K(x,t)\left\{ f(t) + \lambda \sum_\nu \frac{(f\psi_\nu)}{\lambda_\nu - \lambda}\psi_\nu(t)\right\} dt.$$

The right-hand member may be written

$$(107)\quad f(x) + \lambda \sum_\nu \frac{(f\psi_\nu)}{\lambda_\nu - \lambda}\psi_\nu(x) - \lambda \int_a^b K(x,t)f(t)dt$$
$$- \lambda^2 \sum_\nu \frac{(f\psi_\nu)}{\lambda_\nu - \lambda}\int_a^b K(x,t)\psi_\nu(t)dt.$$

But
$$\int_a^b K(x,t)\psi_\nu(t)dt = \frac{\psi_\nu(x)}{\lambda_\nu},$$

therefore (107) may be written

$$f(x) + \lambda \sum_\nu \frac{(f\psi_\nu)}{\lambda_\nu - \lambda}\psi_\nu(x)\left\{1 - \frac{\lambda}{\lambda_\nu}\right\} - \lambda \int_a^b K(x,t)f(t)dt$$

or

$$f(x) + \lambda \sum_\nu \frac{(f\psi_\nu)}{\lambda_\nu}\psi_\nu(x) - \lambda \int_a^b K(x,t)f(t)dt.$$

By (99) this last expression reduces to $f(x)$ and hence

$$u(x) - \lambda \int_a^b K(x,t)u(t)dt = f(x).$$

We have thus proved the following theorem:

Theorem XVIII.—*If* $f(x)$ *is continuous and* λ *is not a characteristic constant, then*

$$u(x) = f(x) + \lambda \int_a^b K(x, t)u(t)dt$$

has one and only one continuous solution $u(x)$ *given by*

$$u(x) = f(x) + \lambda \sum_\nu \left[\frac{1}{\lambda_\nu - \lambda} \int_a^b f(t)\psi_\nu(t)dt \right] \psi_\nu(x),$$

the series being absolutely and uniformly convergent.

70. Schmidt's Solution of the Non-homogeneous Integral Equation When λ Is a Characteristic Constant.—Let us suppose that λ *is* a characteristic constant, for example, $\lambda = \lambda_1$ of index q, so that

$$\lambda = \lambda_1 = \lambda_2 = \ . \ . \ . = \lambda_q.$$

A) As before, we find that

$$(\lambda_\nu - \lambda)(u\psi_\nu) = \lambda_\nu(f\psi_\nu).$$

Now, since $\lambda_\nu = \lambda$, $\nu = 1, 2, \ . \ . \ . \ , q$,
we have

$$(f\psi_\nu) = 0, \quad \nu = 1, 2, \ . \ . \ . \ , q. \tag{108}$$

We have thus q *necessary* conditions on f. If $\nu > q$, then $\lambda_\nu - \lambda \neq 0$ and

$$(u\psi_\nu) = \frac{\lambda_\nu(f\psi_\nu)}{\lambda_\nu - \lambda} \text{ as before.}$$

Now v(x) is expressible as before by the formula (102), but for $\nu = 1, \ . \ . \ . \ , q$ the constant coefficients $\frac{(u\psi_\nu)}{\lambda_\nu}$ are as yet undetermined. So we write

$$\mathrm{v}(x) = C_1\psi_1(x) + \ . \ . \ . + C_q\psi_q(x) + \sum_{\nu=q+1}^{\infty} \frac{(f\psi_\nu)}{\lambda_\nu - \lambda}\psi_\nu(x)$$

whence

$$(109) \quad u(x) = f(x) + C_1\psi_1(x) + \cdots + C_q\psi_q(x) + \lambda\sum_{\nu=q+1}^{\infty} \frac{(f\psi_\nu)}{\lambda_\nu - \lambda}\psi_\nu(x).$$

B) That (109) converges is seen as before. To show that (109) satisfies (100) we first show that

$$u_o(x) = f(x) + \lambda\sum_{\nu=q+1}^{\infty} \frac{(f\psi_\nu)}{\lambda_\nu - \lambda}\psi_\nu(x)$$

satisfies (100). For this proceed as before and we find, after some reductions,

$$u_o(x) - \lambda\int_a^b K(x, t)u_o(t)dt = f(x) + \lambda\sum_{\nu=q+1}^{\infty} \frac{(f\psi_\nu)}{\lambda_\nu}\psi_\nu(x) - \lambda\sum_{\nu=1}^{\infty} \frac{(f\psi_\nu)}{\lambda_\nu}\psi_\nu(x).$$

But $(f\psi_\nu) = 0$ for $\nu = 1, 2, \ldots, q$, by (108) and thus we get

$$(110) \qquad u_o(x) - \lambda\int_a^b K(x, t)u_o(t)dt = f(x).$$

It remains to show that

$$u(x) = u_o(x) + C_1\psi_1(x) + \cdots + C_q\psi_q(x) \equiv u_o(x) + \varphi(x)$$

satisfies (100). Now

$$u(x) - \lambda\int_a^b K(x, t)u(t)dt = u_o(x) - \lambda\int_a^b K(x, t)u_o(t)dt + \varphi(x) - \lambda\int_a^b K(x, t)\varphi(t)dt = f(x) \text{ by (110)},$$

for

$$\varphi(x) - \lambda\int_a^b K(x, t)\varphi(t)dt = 0,$$

since

$$\psi_\alpha(x) - \lambda\int_a^b K(x, t)\psi_\alpha(t)dt = 0 \ (\alpha = 1, \ldots, q).$$

We have thus proved the following theorem:

Theorem XIX.—*If* $\lambda = \lambda_1$ *is a characteristic constant of index* q, *then*

$$u(x) = f(x) + \lambda_1 \int_a^b K(x, t)u(t)dt$$

has in general no solution. Solutions exist only if

$$(f\psi_\alpha) = 0, \ (\alpha = 1, \ . \ . \ . \ , q).$$

Then there are q^∞ *solutions given by*

$$(111) \quad u(x) = f(x) + \lambda \sum_{\nu=q+1}^{\infty} \left[\frac{1}{\lambda_\nu - \lambda} \int_a^b f(t)\psi_\nu(t)dt \right] \psi_\nu(x) + C_1\psi_1(x) + \ . \ . \ . \ + C_q\psi_q(x),$$

where $C_1, \ . \ . \ . \ , C_q$ *are arbitrary constants.*

71. Remarks on Obtaining a Solution.—To obtain the solution of any numerical problem, one must compute $D(\lambda)$ in order to obtain λ_ν. To obtain the ψ_ν one must solve the corresponding homogeneous equation. Having obtained the ψ_ν and the λ_ν, one writes down the solution at once by direct substitution in the proper formula, (104) or (111).

In general, $D(\lambda)$ is an infinite series given by (5) in Chap. III. However, for many simple kernels all but a few of the first terms vanish identically and we obtain $D(\lambda)$ as a polynomial in λ. By the method of §25, Chap. III, $D(\lambda)$ can be obtained in finite form as a polynomial in λ for every kernel $K(x, t)$ which is a polynomial in x and t.

The ψ_ν can be computed from the formula for $\varphi_\alpha(x)$ given in Theorem VIII, §21, Chap. III. By the method of §25, Chap. III, ψ_ν can be obtained in finite form for every kernel $K(x, t)$ which is a polynomial in x and t.

EXERCISES

Compute $D(\lambda)$ and obtain the characteristic constants λ_ν for the following symmetric kernels for the specified interval $[ab]$.

Ans.

1. $K(x, t) = 1$, [01]. $\quad D(\lambda) = 1 - \lambda.$
2. $K(x, t) = -1$, [01]. $\quad D(\lambda) = 1 + \lambda.$
3. $K(x, t) = xt$, [01]. $\quad D(\lambda) = 1 - \frac{\lambda}{3}.$
4. $K(x, t) = \sin x \sin t$ $[0, 2\pi]$. $\quad D(\lambda) = 1 - \pi\lambda.$
5. $K(x, t) = e^x \cdot e^t [0, \log_e 2]$. $\quad D(\lambda) = 1 - \frac{3}{2}\lambda.$
6. $K(x, t) = x + t$, [01]. $\quad D(\lambda) = 1 - \lambda - \frac{\lambda^2}{12}.$
7. $K(x, t) = x^2 + t^2$, [01]. $\quad D(\lambda) = 1 - \frac{2}{3}\lambda - \frac{4}{45}\lambda^2.$
8. $K(x, t) = x^2t + xt^2$, [01]. $\quad D(\lambda) = 1 - \frac{\lambda}{2} - \frac{1}{240}\lambda^2.$
9. $K(x, t) = x^2 + xt + t^2$, [01]. $\quad D(\lambda) = 1 - \lambda - \frac{7}{60}\lambda^2 + \frac{1}{2{,}160}\lambda^3.$

Solve the following equations:

10. $u(x) = \int_0^1 u(t)dt.$ $\quad u(x) = C.$

11. $u(x) = x + \lambda \int_0^1 u(t)dt, \; (\lambda \neq 1).$ $\quad u(x) = x + \frac{\lambda}{2(1 - \lambda)}.$

12. $u(x) = \frac{1}{2} - x + \int_0^1 u(t)dt.$ $\quad u(x) = \frac{1}{2} - x + C.$

13. $u(x) = \lambda \int_0^1 (x + t)u(t)dt$ for $\lambda = -6 \pm 4\sqrt{3}.$

Ans. $u_1(x) = C_1(1 + \sqrt{3}x)$ for $\lambda = \lambda_1 = -6 + 4\sqrt{3}.$
$u_2(x) = C_2(1 - \sqrt{3}x)$ for $\lambda = \lambda_2 = -6 - 4\sqrt{3}.$

14. $u(x) = x + \lambda \int^1 (x + t)u(t)dt, \; \lambda \neq \lambda_1, \lambda_2.$

$$\textit{Ans. } u(x) = x - \frac{\frac{\lambda}{2}x - 2 - \sqrt{3}}{\lambda + 6 + 4\sqrt{3}} - \frac{\frac{\lambda}{2}x + 2 - \sqrt{3}}{\lambda + 6 - 4\sqrt{3}}$$

$$\text{or } u(x) = \frac{(6\lambda - 12)x - 4\lambda}{\lambda^2 + 12\lambda - 12}.$$

15. $u(x) = (1 - \sqrt{3}x) + (-6 + 4\sqrt{3}) \int_0^1 (x + t)u(t)dt.$

$$\textit{Ans. } u(x) = (1 - \sqrt{3}x) + B(1 + \sqrt{3}x) - \left(1 + \frac{3}{2}x\right).$$

16. $u(x) = (1 + \sqrt{3}x) + (-6 - 4\sqrt{3}) \int_0^1 (x + t)u(t)dt.$

$$\textit{Ans. } u(x) = (1 + \sqrt{3}x) + B(1 - \sqrt{3}x) - \left(1 + \frac{3}{2}x\right).$$

CHAPTER VI

APPLICATIONS OF THE HILBERT-SCHMIDT THEORY

I. Boundary Problems for Ordinary Linear Differential Equations

72. Introductory Remarks.—*a) Formulation of the Problem.*—We have previously considered the boundary problem

$$\frac{d^2u}{dx^2} + \lambda u = 0, \; u(0) = 0, \; u(1) = 0.$$

We shall now discuss the general type of boundary problem of which this is a special case, namely, we take the general homogeneous linear differential equation of the second order with a parameter λ contained linearly in the coefficients of u:

$$P(x)\frac{d^2u}{dx^2} + Q(x)\frac{du}{dx} + \Big[R(x) + \lambda S(x)\Big]u = 0.$$

We first reduce this equation to a self-adjoint linear differential equation by multiplying it by $\dfrac{1}{P(x)}e^{\int \frac{Q(x)}{P(x)}dx}$

Putting $p = e^{\int \frac{Q}{P}dx}$, we obtain

$$p\left(u'' + \frac{Q}{P}u' + \frac{R + \lambda S}{P}u\right) = 0,$$

which may be written

$$(1) \qquad pu'' + p'u' + (q + \lambda g)u = 0$$

or

$$\frac{d}{dx}\left(p\frac{du}{dx}\right) + (q + \lambda g)u = 0.$$

If we make use of the permanent notation

$$L(u) = \frac{d}{dx}\left(p\frac{du}{dx}\right) + qu, \tag{2}$$

equation (1) may be written in the form

$$L(u) + \lambda g(u) = 0. \tag{3}$$

We use also more general boundary conditions, and for the general interval $\left[ab\right]$:

$$\begin{aligned} R_0(u) &\equiv Au(a) + Bu'(a) = 0 \\ R_1(u) &\equiv Cu(b) + Du'(b) = 0, \end{aligned} \tag{4}$$

where A, B, C, D are given constants, of which A and B are not both zero, and C and D are not both zero.

Our problem then is to determine all solutions of class C'' of the differential equation (3) which satisfy the boundary conditions (4).

For this generalized boundary problem we make the hypotheses:

(H_1) *p is of class C', $p \neq 0$ on $[ab]$*

q and g are of class C on $[ab]$.

Then (1) has two linearly independent solutions of class C'' on $\left[ab\right]$: u_1 and u_2. Any other solution of (1) of class C'' is linearly expressible in terms of u_1 and u_2:

$$u = C_1u_1 + C_2u_2.$$

The boundary problem has the trivial solution $u \equiv 0$. Every value of λ for which our boundary problem has a non-trivial solution is called a characteristic constant and those solutions of class C'', non-trivial, which exist for these values of λ are called fundamental functions of the boundary problem.

To the hypothesis (H_1) we add for the present a second hypothesis (H_2) which will be dropped later:

(H_2) $\lambda = 0$ *is not a characteristic constant.*

That is, $L(u) = 0,\ R_0(u) = 0,\ R_1(u) = 0$

has no solution of class C'' other than $u \equiv 0$.

b) Green's Formula.—From our hypothesis (H_1) we obtain the following theorem:

Theorem I.—*If* $u(x)$ *and* $\mathrm{v}(x)$ *are any two functions of* x *of class* C'', *then*

$$uL(\mathrm{v}) - \mathrm{v}L(u) = \frac{d}{dx}p(u\mathrm{v}' - \mathrm{v}u'). \tag{5}$$

Proof.—Given two functions $u(x)$, $\mathrm{v}(x)$ of class C'', form

$$uL(\mathrm{v}) - \mathrm{v}L(u) = \begin{vmatrix} u & pu'' + p'u' + qu \\ \mathrm{v} & p\mathrm{v}'' + p'\mathrm{v}' + q\mathrm{v} \end{vmatrix}$$

$$= \begin{vmatrix} u & pu'' + p'u' \\ \mathrm{v} & p\mathrm{v}'' + p'\mathrm{v}' \end{vmatrix}$$

$$= p(u\mathrm{v}'' - \mathrm{v}u'') + p'(u\mathrm{v}' - \mathrm{v}u')$$

$$= \frac{d}{dx}p(u\mathrm{v}' - \mathrm{v}u').$$

Corollary.—*If* u *and* v *are of class* C'' *and* $L(u) = 0$, $L(\mathrm{v}) = 0$, then

$$p(u\mathrm{v}' - \mathrm{v}u') = \text{constant} \neq 0, \tag{6}$$

if u *and* v *are independent.*

For if $C = 0$, then $u\mathrm{v}' - \mathrm{v}u' \equiv 0$ and

$$\frac{u\mathrm{v}' - \mathrm{v}u'}{u^2} \equiv 0.$$

Whence $$\frac{d}{dx}\left(\frac{\mathrm{v}}{u}\right) \equiv 0,$$

and, therefore, $\mathrm{v} = Cu$ and v and u would be linearly dependent, contrary to hypothesis.

c) Consequences of Hypothesis (H_2).—We now establish the following lemmas:

Lemma I.—If u_1, u_2 *are of class* C'' *and* $L(u_1) = 0$, $L(u_2) = 0$, *and* u_1, u_2 *are linearly independent, then*

$$R_0(u_1),\ R_0(u_2) \text{ are not both zero and also}$$
$$R_1(u_1),\ R_1(u_2) \text{ are not both zero.}$$

Proof.—The general solution of class C'' of $L(u) = 0$ is

$$u = c_1u_1 + c_2u_2.$$

Now
$$R_0(u) = c_1R_0(u_1) + c_2R_0(u_2).$$

Therefore, if the lemma is not true, then

$$R_0(u) = 0 \text{ for every } c_1, c_2.$$

But

$$R_1(u) = c_1R_1(u_1) + c_2R_1(u_2),$$

and, if c_1, c_2 are properly chosen, we can make $R_1(u) = 0$. But then we would have

$$L(u) = 0,\ R_0(u) = 0,\ R_1(u) = 0,$$

which is contrary to hypothesis (H_2). Therefore not both $R_0(u_1)$, $R_0(u_2)$ can vanish at once. Likewise, we show that $R_1(u_1)$, $R_1(u_2)$ cannot both vanish simultaneously.

Lemma II.—There exist functions u *and* v *of class* C'', *determined except for a constant factor, such that* $L(u) = 0$, $R_0(u) = 0$, $L(\mathrm{v}) = 0$, $R_1(\mathrm{v}) = 0$. u and v *are linearly independent and the constant factors can be determined so that* $p(u\mathrm{v}' - \mathrm{v}u') = -1$.

Proof.—Let u_1, u_2 be two linearly independent solutions of $L(u) = 0$. Let $u = c_1u_1 + c_2u_2$ and determine c_1 and c_2 so that $R_0(u) = c_1R_0(u_1) + c_2R_0(u_2) = 0$. Now $R_0(u_1)$, $R_0(u_2)$ cannot both be zero. Therefore

$$c_1 = \rho_0R_0(u_2) \text{ and } c_2 = -\rho_0R_0(u_1),$$

and hence

$$u = \rho_0\left[u_1R_0(u_2) - u_2R_0(u_1)\right].$$

This is the most general expression for a function u for which $L(u) = 0$, $R_0(u) = 0$.

In like manner, we show that the most general expression for a function v for which

$$L(\mathrm{v}) = 0,\ R_1(\mathrm{v}) = 0$$

is

$$\mathrm{v} = \rho_1 \Big[\mathrm{v}_1 R_1(\mathrm{v}_2) - \mathrm{v}_2 R_1(\mathrm{v}_1) \Big].$$

Suppose u and v were not independent, then

$$\mathrm{v} = Cu,\ C \neq 0,$$

whence

$$R_1(u) = \frac{1}{C} R_1(\mathrm{v}) = 0.$$

But by hypothesis $R_0(u) = 0$. This is in contradiction to our hypothesis (H_2).

According to Green's formula

$$p(u\mathrm{v}' - \mathrm{v}u') = C \neq 0;$$

hence the constant factors ρ_0, ρ_1 can be determined so that

$$p(u\mathrm{v}' - \mathrm{v}u') = -1.$$

73. Construction of Green's Function.—Let us take between a and b a fixed value t: $a < t < b$. Then the following theorem holds:

Theorem II.—*For our boundary problem there exists under the assumptions (H_1) and (H_2) one and but one function $K(x, t)$, which, as a function of x, has the following properties:*

A) K is continuous on $\Big[ab\Big]$.

B) K is of class C'' on each of its subintervals $\Big[at\Big]\Big[tb\Big]$, and $L(K) = 0$ on each subinterval.

C) $R_0(K) = 0$, $R_1(K) = 0$.

D) $K'(x, t - 0) - K'(x, t + 0) = \dfrac{1}{p(t)},$

where $$K'(x, t) = \frac{\partial}{\partial x} K(x, t).$$

This function $K(x, t)$ is called *Green's function*, belonging to the boundary problem.

Proof.—If $K(x, t)$ is to satisfy $L(K) = 0$, it must be linearly expressible in terms of the two functions u and v the existence of which have been shown in Lemma II of §72:

$$K(x, t) = \begin{cases} A_0 u + B_0 \mathrm{v}, & [at] \\ A_1 u + B_1 \mathrm{v}, & [tb]. \end{cases}$$

Now demand that $K(x, t)$ satisfy C):

$$R_0(K) = R_0(A_0 u + B_0 \mathrm{v}) = A_0 R_0(u) + B_0 R_0(\mathrm{v}) = 0.$$
$$R_1(K) = R_1(A_1 u + B_1 \mathrm{v}) = A_1 R_1(u) + B_1 R_1(\mathrm{v}) = 0.$$

But $R_0(u) = 0$, and not both $R_0(u)$, $R_0(\mathrm{v})$ can vanish simultaneously, whence $B_0 = 0$. Also, $R_1(\mathrm{v}) = 0$, and not both $R_1(u)$, $R_1(\mathrm{v})$ can vanish simultaneously, whence $A_1 = 0$. Therefore

$$K(x, t) = \begin{cases} A_0 u, & [at] \\ B_1 \mathrm{v}, & [tb]. \end{cases}$$

We now impose the condition A):

$$A_0 u(t) = B_1 \mathrm{v}(t),$$

whence

$$A_0 = \rho \mathrm{v}(t), \; B_1 = \rho u(t),$$

and

$$K(x, t) = \begin{cases} \rho \mathrm{v}(t) u(x), & [at] \\ \rho u(t) \mathrm{v}(x), & [tb]. \end{cases}$$

Now impose the condition D):

$$K'(t - 0) - K'(t + 0) = \frac{1}{p(t)},$$

that is

$$\rho \mathrm{v}(t)u'(t) - \rho u(t)\mathrm{v}'(t) = \frac{1}{p(t)}.$$

But $$p(u\mathrm{v}' - \mathrm{v}u') = -1.$$

Therefore $$\rho\Big[\mathrm{v}(t)u'(t) - u(t)\mathrm{v}'(t)\Big] = \frac{\rho}{p(t)},$$

whence $$\rho = +1.$$

Therefore

$$(7) \qquad K(x, t) = \begin{cases} \mathrm{v}(t)u(x) \equiv K_0(x, t), \; a \leqq x \leqq t \\ u(t)\mathrm{v}(x) \equiv K_1(x, t), \; t \leqq x \leqq b. \end{cases}$$

This is *Green's function* and it is uniquely determined.

Corollary.—Green's function is symmetric. That is

$$K(x, t) = K(t, x).$$

To show this, compute $K(z_1, z_2)$ and $K(z_2, z_1)$ where

$$0 \leqq z_1 < z_2 \leqq b.$$

$$K(z_1, z_2) = \mathrm{v}(z_2)u(z_1),$$

for here we identify z_1 and x, z_2 and t; and, since $z_1 < z_2$, we use the first expression for K, namely, $K_o(x, t)$. Similarly

$$K(z_2, z_1) = u(z_1)\mathrm{v}(z_2).$$

A comparison shows that

$$K(z_1, z_2) = K(z_2, z_1),$$

that is, $$K(x, t) = K(t, x).$$

74. Equivalence Between the Boundary Problem and a Homogeneous Linear Integral Equation. *a) Hilbert's Fundamental Theorem.*—Under the assumptions (H_1), (H_2) let us consider the non-homogeneous linear differential equation

$$L(u) + f = 0,$$

where f is supposed to be continuous on $\Big[ab\Big]$. We can then prove the following theorem:

Theorem III A.—*If F is of class C'' and $L(F) + f = 0$, $R_0(F) = 0$, $R_1(F) = 0$, then*

$$F(x) = \int_a^b K(x, t)f(t)dt.$$

Proof.—We have

$$L(F) = -f \text{ on } \Big[ab\Big], \text{ and}$$

$$L(K) = 0 \quad \text{on } \Big[at\Big], \Big[tb\Big] \text{ separately.}$$

Multiply the first by $-K$ and the last by F and add. We obtain

$$(8) \quad FL(K) - KL(F) = Kf \text{ on } \Big[at\Big], \Big[tb\Big] \text{ separately.}$$

Apply Green's formula to the left-hand side of (8). We obtain

$$(9) \qquad \frac{d}{dx}p(FK' - KF') = Kf.$$

Integrate (9) from a to t and from t to b. We have

$$\Big[p(FK' - KF')\Big]_a^{t-0} = \int_a^{t-0} K(x, t)f(x)dx$$

$$\Big[p(FK' - KF')\Big]_{t+0}^{b} = \int_{t+0}^{b} K(x, t)f(x)dx.$$

Add the last two equations. We obtain

$$(10) \quad \Big[p(FK' - KF')\Big]_{t+0}^{t-0} - \Big[p(FK' - KF')\Big]^{x=a} + \Big[p(FK' - KF')\Big]^{x=b} = \int_a^b K(x, t)f(x)dx,$$

since $K(x, t)$ is continuous at $x = t$. Now p, F, F', and K are continuous at $x = t$.

Hence $$-KF'\Big]_{t+0}^{t-0} = 0.$$

While

$$p(FK')\Big]_{t+0}^{t-0} = p(t-0)F(t-0)K'(t-0)$$
$$- p(t+0)\,F(t+0)K'(t+0)$$
$$= p(t)F(t)\Big[K'(t-0) - K'(t+0)\Big]$$

which by $D)$

$$= p(t)F(t)\frac{1}{p(t)}.$$

Now write the second term in (10) as a determinant

$$p(a)\begin{vmatrix} F(a) & F'(a) \\ K(a) & K'(a) \end{vmatrix}.$$

If $A \neq 0$, then

$$\Big[p(FK' - F'K)\Big]^{x=a} = \frac{p(a)}{A}\begin{vmatrix} AF(a) + BF'(a) & F'(a) \\ AK(a) + BK'(a) & K'(a) \end{vmatrix}$$
$$\equiv \frac{p(a)}{A}\begin{vmatrix} R_0(F) & F'(a) \\ R_0(K) & K'(a) \end{vmatrix} = 0,$$

since $\quad R_0(F) = 0,\ R_0(K) = 0.$

If $A = 0$, then $B \neq 0$ and, as before,

$$\Big[p(FK' - KF')\Big]^{x=a} = \begin{vmatrix} F(a) & R_0(F) \\ K(a) & R_0(K) \end{vmatrix}\frac{p(a)}{B} = 0.$$

Likewise $\quad \Big[p(FK' - KF')\Big]^{x=b} = 0.$

Therefore, $\quad F(t) = \int_a^b K(x, t)f(x)dx.$

Whence, interchanging x and t, on account of the symmetry of $K(x, t)$, we have

$$(11) \qquad F(x) = \int_a^b K(x, t)f(t)dt.$$

Corollary.—If $F(x)$ is of class C'' and $R_0(F) = 0, R_1(F) = 0$, and $f = -L(F)$, then

$$(12) \qquad F(x) = \int_a^b K(x, t)\Big[-L(F)\Big]dt.$$

Equation (11) is an integral equation of the *first* kind, with f as the function to be determined. Equation (12) shows us that (11) has a solution:

$$f = -L(F).$$

Theorem IIIB.—*If* $F(x) = \int_a^b K(x, t)f(t)dt$, *then* F *is of class* C'', $L(F) + f = 0$, $R_0(F) = 0$, $R_1(F) = 0$.

Proof.—In order to form the derivatives of $F(x)$, we break the integral into two parts, inasmuch as K' is not continuous for $x = t$.

$$(13) \qquad F(x) = \int_a^x K_1(x, t)f(t)dt + \int_x^b K_0(x, t)f(t)dt$$

$$F'(x) = \int_a^x K_1'(x, t)f(t)dt + \int_x^b K_0'(x, t)f(t)dt + K_1(x, x)f(x) - K_0(x, x)f(x).$$

But $K_1(x, x) - K_0(x, x) = 0$, since K is continuous. Therefore

$$(14) \quad F'(x) = \int_a^x K_1'(x, t)f(t)dt + \int_x^b K_0'(x, t)f(t)dt.$$

$$F''(x) = \int_a^x K_1''(x, t)f(t)dt + \int_x^b K_0''(x, t)f(t)dt + K_1'(x, x)f(x) - K_0'(x, x)f(x).$$

But $\qquad K_1'(x, x) - K_0'(x, x) = \dfrac{-1}{p(x)}.$

Therefore

$$(15) \quad F''(x) = \int_a^x K_1''(x, t)f(t)dt + \int_x^b K_0''(x, t)f(t)dt - \frac{f(x)}{p(x)}.$$

Multiply (13), (14), (15) by $q(x)$, $p'(x)$, $p(x)$ respectively and add. We get

$$L(F) = \int_a^x L(K_1)f(t)dt + \int_x^b L(K_0)f(t)dt - f(x).$$

But $L(K_1) = 0$ and $L(K_0) = 0$, therefore

$$L(F) = -f(x). \tag{16}$$

Now form $R_0(F)$.

$$R_0(F) = AF(a) + BF'(a)$$
$$= \int_a^b R_0(K_0)f(t)dt = 0,$$

since $R_0(K_0) = 0$. Similarly, we show that $R_1(F) = 0$.

That $F(x)$ is of class C'' follows from (13), (14), and (15), if we substitute for K_0, K_1 their explicit expressions from (7).

We combine Theorems III*A* and III*B* into the following, which is Hilbert's third fundamental theorem:

Theorem III.—*If $f(x)$ is continuous, then*

$$F(x) = \int_a^b K(x, t)f(t)dt$$

implies and is implied by $F(x)$ is of class C'',

$$L(F) + f = 0,\ R_0(F) = 0, \text{ and } R_1(F) = 0.$$

b) Equivalence of Boundary Problem and Integral Equation.—If in Theorem III we put

$$f(x) = \lambda g(x)u(x),\ F(x) = u(x)$$

we obtain the theorem:

Theorem IV.—*If $u(x)$ is continuous, then*

$$u(x) = \lambda \int_a^b K(x, t)g(t)u(t)dt \tag{17}$$

implies and is implied by $u(x)$ is of class C'',

$$L(u) + \lambda gu = 0,\ R_0(u) = 0, \text{ and } R_1(u) = 0.$$

For λ is constant and by (H_1) g is continuous, therefore the hypothesis that $f(x)$ is continuous becomes $u(x)$ is continuous.

We notice that (17) is a homogeneous linear integral equation of the second kind with, in general, an unsymmetric kernel:

$$H(x, t) \equiv K(x, t)g(t). \tag{18}$$

We remark further that the condition that $u(x)$ be of class C'' carries with it the condition that $u(x)$ is of class C. Thus we drop the explicit statement that $u(x)$ be continuous and obtain from Theorem IV the two following theorems:

Theorem IVA.—*The conditions*

a) $u(x)$ is of class C''.

b) $L(u) + \lambda g u = 0,\ R_0(u) = 0,\ R_1(u) = 0$

imply that

$$u(x) = \lambda \int_a^b K(x, t) g(t) u(t) dt.$$

Theorem IVB.—*The conditions*

a) $u(x)$ is continuous.

b) $u(x) = \lambda \int_a^b K(x, t) g(t) u(t) dt$

imply that

a) $u(x)$ is of class C''.

b) $L(u) + \lambda g u = 0,\ R_0(u) = 0,\ R_1(u) = 0.$

Theorems IV*A* and IV*B* establish the equivalence of the boundary problem and of the integral equation. Hence,

1) If λ_0 is a characteristic constant of the boundary problem, then λ_0 is a characteristic constant of the integral equation and *vice versa*.

2) If $u(x)$ is a fundamental function for the boundary problem, belonging to λ_0, then $u(x)$ is a fundamental function, belonging to λ_0, for the integral equation and *vice versa*.

We remark that for this particular kernel (18), the fundamental functions are of class C''.

75. Special Case g(x) ≡ 1.—For the special case $g(x) \equiv 1$, the equivalence between the boundary problem and the integral equation becomes:

$$(19)\qquad \begin{cases} L(u) + \lambda u = 0,\ R_0(u) = 0,\ R_1(u) = 0 \\ \text{implies and is implied by} \\ u(x) = \lambda \int_a^b K(x, t) u(t) dt. \end{cases}$$

The kernel of the integral equation in (19) is now symmetric and thus the results of both the Fredholm and the Hilbert-Schmidt theory apply.

Let us state some of the results of the Hilbert-Schmidt theory:

1) There exists at least one characteristic constant.
2) All of the characteristic constants are real.
3) The index $q =$ the multiplicity r.
4) There exists a complete normalized orthogonal system of fundamental functions $\left\{ \psi_\nu(x) \right\}$ with corresponding characteristic constants $\left\{ \lambda_\nu \right\}$, $\nu = 1, 2, \ldots$.
5) If $\sum_\nu \frac{\psi_\nu(x)\psi_\nu(t)}{\lambda_\nu}$ is uniformly convergent on R, then

$$\sum_\nu \frac{\psi_\nu(x)\psi_\nu(t)}{\lambda_\nu} = K(x, t). \tag{20}$$

6) If $u(x)$ is expressible in the form

$$u(x) = \lambda \int_a^b K(x, t)u(t)dt$$

and if $u(x)$ is continuous, then

$$u(x) = \sum_\nu C_\nu \psi_\nu(x), \tag{21}$$

where $C_\nu = (u\psi_\nu)$, and the series is absolutely and uniformly convergent on $\left[ab \right]$.

For the special case under consideration, each of these six statements can be made for the boundary problem on account of the equivalence established.

Further results follow, for the kernel under consideration, besides satisfying the conditions $K(x, t)$, is continuous, $\not\equiv 0$, real, and symmetric of the Hilbert-Schmidt theory, satisfies now, in addition, the four conditions $A)$ $B)$ $C)$ $D)$ of §73.

Theorem V.—*There exists always an infinite number of characteristic constants* $\left\{ \lambda\nu \right\}$ *and a corresponding complete normalized orthogonal system of fundamental functions* $\left\{ \psi_\nu(x) \right\}$.

Proof.—Suppose that there is a finite number of characteristic constants. Then there are a finite number of fundamental functions:

$$\psi_1(x), \ . \ . \ . \ , \psi_m(x)$$
$$\lambda_1, \ . \ . \ . \ , \lambda_m.$$

For these the bilinear formula (20) holds. But $\psi_\nu(x)$ is of class C'' on R and, therefore, $K(x, t)$ would be of class C'' on R, which contradicts D), namely

$$K'(x, t-0) - K'(x, t+0) = \frac{1}{p(t)}.$$

Theorem VI.—*If* $u(x)$ *is of class* C'' *and* $R_0(u) = 0$,

$$R_1(u) = 0,$$

then

$$u(x) = \sum_\nu C_\nu \psi_\nu(x)$$

where $C_\nu = (u\psi_\nu)$, *and the series is absolutely and uniformly convergent.*

Proof.—From the corollary to Theorem IIIA

$$u(x) = \int_a^b K(x, t)\left[- L(u) \right] dx.$$

But $L(u)$ is continuous and, therefore, from (21)

$$u(x) = \sum_\nu C_\nu \psi_\nu(x).$$

Theorem VII.—*For every characteristic constant* λ *the index* q *is unity:* $q = 1$.

Proof.—Suppose $q > 2$, then at least three of the characteristic constants are equal, say, $\lambda_1 = \lambda_2 = \lambda_3$. Let the corresponding fundamental functions be $\psi_1(x)$, $\psi_2(x)$, $\psi_3(x)$. We would then have three different independent solutions for $\lambda = \lambda_1$, of the same integral equation, that is, by the equivalence theorem three different solutions of the same differential equation of the second order. Hence one of them, say, ψ_3, is linearly expressible in terms of the other two:

$$\psi_3(x) = C_1\psi_1 + C_2\psi_2.$$

This is a contradiction. Therefore $q \not> 2$.

Suppose $q = 2$. Then there would be two fundamental functions $\psi_1(x)$, $\psi_2(x)$ belonging to $\lambda = \lambda_1$. Now, by Green's formula,

$$p\Big[\psi_1(x)\psi_2'(x) - \psi_2(x)\psi_1'(x)\Big] = C \neq 0. \tag{22}$$

But

$$\begin{aligned} R_0(\psi_1) &= A\psi_1(a) + B\psi_1'(a) = 0 \\ R_0(\psi_2) &= A\psi_2(a) + B\psi_2'(a) = 0. \end{aligned}$$

These are two linear homogeneous equations for the determination of A, B, not both zero. Hence

$$\psi_1(a)\psi_2'(a) - \psi_2(a)\psi_1'(a) = 0,$$

which contradicts (22), therefore $q \neq 2$. Therefore we have $q = 1$.

We shall now discuss the more special boundary problem in which it follows from the boundary conditions that

$$\Big[puu'\Big]_a^b = 0. \tag{23}$$

For instance, if we have

$$u(a) = 0,\ u(b) = 0, \text{ or } u(a) = 0,\ u'(b) = 0,$$

or $$u'(a) = 0,\ u(b) = 0, \text{ or } u'(a) = 0,\ u'(b) = 0,$$

equation (23) is satisfied.

For this more special boundary problem we have the theorem:

Theorem VIII.—*If, as a consequence of the boundary conditions, we have*

$$\Big[puu'\Big]_a^b = 0,$$

then there exists a $\begin{Bmatrix} \textit{smallest} \\ \textit{largest} \end{Bmatrix}$ *characteristic constant if* $\begin{Bmatrix} p(x) > 0 \\ p(x) < 0 \end{Bmatrix}$.

Proof.—The assumptions $p(x) > 0$, $p(x) < 0$ are in harmony with our previous assumption that $p(x) \neq 0$ on $\Big[ab\Big]$, for then $p(x)$ is always of the same sign. We have previously shown that there are an infinite number of characteristic constants, but, so far as we know, at present they may be infinite in number in both the positive and negative directions. Let $\psi_\nu(x)$ be a fundamental function belonging to λ_ν, then

$$\begin{aligned} &L(\psi_\nu) + \lambda_\nu\psi_\nu = 0, \\ &R_0(\psi_\nu) = 0,\ R_1(\psi_\nu) = 0, \end{aligned} \tag{24}$$

ψ_ν is of class C'', and $\psi_\nu \not\equiv 0$. Multiply both members of (24) by ψ_ν and integrate. We obtain

$$\lambda_\nu = -\int_a^b \psi_\nu(x)L(\psi_\nu)dx \tag{25}$$

since

$$\int_a^b \psi_\nu^2(x)dx = 1.$$

In (25) put the explicit expression for $L(\psi_\nu)$. We have

$$\lambda_\nu = -\int_a^b \left[\psi_\nu \frac{d}{dx}(p\psi_\nu') + q\psi_\nu^2\right]dx,$$

or

$$\lambda_\nu = -\int_a^b \psi_\nu \frac{d}{dx}(p\psi_\nu')dx + \int_a^b (-q)\psi_\nu^2 dx.$$

Integrate the first integral by parts. We find

$$\lambda_\nu = -\left[p\psi_\nu\psi_\nu'\right]_a^b + \int_a^b p\psi_\nu'^2dx + \int_a^b (-q)\psi_\nu^2dx.$$

Now $-q$ is a continuous function on $\left[ab\right]$ and hence has a finite maximum M and a finite minimum m:

$$m \leqq -q \leqq M \text{ on } \left[ab\right].$$

Therefore

$$(26) \qquad \begin{cases} \lambda_\nu \leqq \displaystyle\int_a^b p\psi_\nu'^2dx + M, \text{ and} \\ \lambda_\nu \geqq \displaystyle\int_a^b p\psi_\nu'^2dx + m. \end{cases}$$

If $p > 0$ on $\left[ab\right]$, then the integral in (26) is a positive number and therefore $\lambda_\nu > m$.

If $p < 0$ on $\left[ab\right]$, then the integrand in (26) is a negative number and therefore $\lambda_\nu < M$.

Combining these results with the fact that a finite interval can contain only a finite number of the λ_ν's, we obtain the result:

When $p(x) > 0$, the characteristic constants can be arranged in a sequence increasing towards plus infinity:

$$p(x) > 0; \lambda_1 < \lambda_2 < \lambda_3 < \ . \ . \ . \ < \lambda_\nu < \ . \ . \ .$$

and when $p(x) < 0$, they may be arranged in a sequence decreasing towards minus infinity:

$$p(x) < 0; \lambda_1 > \lambda_2 > \lambda_3 > \ . \ . \ . \ > \lambda_\nu > \ . \ . \ .$$

Corollary.—If $p(x) > 0$ and $q \leqq 0$ on $\left[ab\right]$, then all of the characteristic constants are positive; if $p(x) < 0$ and $q \geqq 0$ on $\left[ab\right]$, then all of the characteristic constants are negative.

For if $-q \geqq 0$, then $m \geqq 0$, so that, from (26), since $p > 0$, we get

$$m + (a \text{ positive quantity}) \leqq \lambda_\nu.$$

Therefore $$\lambda_\nu > 0.$$

If $-q \leqq 0$, then $M \leqq 0$, so that, from (26), since $p < 0$, we obtain

$$\lambda_\nu \leqq M + (\text{a negative quantity not zero}).$$

Therefore $$\lambda_\nu < 0.$$

Definition.—A real symmetric kernel $K(x, t)$ is said to be **closed** *if there exists no continuous function $h(x)$ other than $h(x) \equiv 0$, for which*

$$\int_a^b K(x, t)h(t)dt = 0, \ (x).$$

Theorem IX.—*Green's function $K(x, t)$ for our boundary problem is always closed.*

Proof.—We make use of Hilbert's third fundamental theorem, which states the equivalence of the boundary problem and the homogeneous linear integral equation:

If $f(x)$ is continuous, then

$$F(x) = \int_a^b K(x, t)f(t)dt$$

implies and is implied by $F(x)$ is of class C'',

$$L(F) + f = 0, \ R_0(F) = 0, \text{ and } R_1(F) = 0.$$

We apply this theorem for $F(x) \equiv 0$, whence if $f(x)$ is continuous and

$$\int_a^b K(x, t)f(t)dt = 0, \ (x),$$

then $$f = -L(0) \equiv 0.$$

76. Miscellaneous Remarks. *a) The General Case $g(x) \not\equiv 1$.*—In the previous article we considered the special case $g(x) \equiv 1$. We now consider the problem for $g(x) \not\equiv 1$. Hilbert's third fundamental theorem now reads:

If $u(x)$ is continuous, then

$$u(x) = \lambda \int_a^b K(x, t)g(t)u(t)dt$$

implies and is implied by $u(x)$ is of class C'',

$$L(u) + \lambda gu = 0, R_0(u) = 0, R_1(u) = 0.$$

There are two cases to be considered:

Case I.—$g(x) \neq 0$ *on* $\left[ab\right]$. This case can be reduced to the case $g(x) \equiv 1$. Since $g(x) \neq 0$ on $\left[ab\right]$, we have either $g(x) > 0$ or $g(x) < 0$ on $\left[ab\right]$. Consider the first case. Multiply both sides of

$$u(x) = \lambda \int_a^b K(x, t)g(t)u(t)dt \tag{27}$$

by $\sqrt{g(x)}$, and put

$$\bar{u}(x) = \sqrt{g(x)}\cdot u(x).$$
$$\bar{K}(x, t) = K(x, t)\sqrt{g(x)}\sqrt{g(t)}.$$

We then obtain

$$\bar{u}(x) = \lambda \int_a^b \bar{K}(x, t)\bar{u}(t)dt. \tag{28}$$

This is the reduction desired, for $\bar{K}(x, t)$ is symmetric.

If $g(x) < 0$ on $\left[ab\right]$, multiply both sides of (27) by $\sqrt{-g(x)}$, and put

$$\bar{u}(x) = \sqrt{-g(x)}\cdot u(x).$$
$$\bar{K}(x, t) = K(x, t)\sqrt{-g(x)}\sqrt{-g(t)}.$$

We obtain

$$\bar{u}(x) = -\lambda \int_a^b \bar{K}(x, t)\bar{u}(t)dt,$$

which is again the reduction desired.

Case II.—$g(x)$ vanishes at some point of $\left[ab\right]$. This case has been treated by Hilbert ["Gott. Nach.," 5, p. 462, 1906] for functions $K(x, t)$ which are *definite*, by means of the theory of quadratic forms with infinitely many variables. J. Marty [*Comptes Rendus*, vol. 150, p. 515; *Ibid.*, p. 605, 1910] has reached the same results without the use of quadratic forms of infinitely many variables.

Definition.—A real symmetric[1] *kernel $K(x, t)$ is said to be definite if no continuous function exists, other than $h(x) \equiv 0$, for which*

$$\int_a^b \int_a^b K(x, t)h(x)h(t)dxdt = 0.$$

The name "definite" has been used on account of the analogy with a *definite* quadratic form $\sum K_{ij}y_iy_j$ which vanishes only when all of the y's vanish.

Integral equations of the form

$$u(x) = f(x) + \lambda \int_a^b K(x, t)g(t)u(t)dt \tag{29}$$

with $K(x, t)$ *definite* have been called by Hilbert *polar* integral equations or integral equations of the third kind. If we multiply the equation (29) by $g(x)$ and put

$$\bar{f}(x) = f(x)g(x)$$
$$\bar{K}(x, t) = K(x, t)g(x)g(t),$$

we obtain the equation in the form in which Hilbert considered it:

$$g(x)u(x) = \bar{f}(x) + \lambda \int_a^b \bar{K}(x, t)u(t)dt.$$

[1] For a discussion of this and other special kernels consult LALESCO, T., "Théorie Des Équations Intégrales," pp. 64*ff*.

Definition.—A kernel $H(x, t)$ is said to be symmetrizable if there exists a definite symmetric kernel $G(x, t)$ such that either

$$K_1(x, t) = \int_a^b G(x, s)H(s, t)ds$$

or

$$K_2(x, t) = \int_a^b H(x, s)G(s, t)ds$$

is symmetrical.

In this instance the kernel

$$H(x, t) = K(x, t)g(t)$$

is symmetrizable if $K(x, t)$ is definite, for

$$K_2(x, t) = \int_a^b K(x, s)g(s)K(s, t)ds$$

is symmetric, since

$$\begin{aligned} K_2(t, x) &= \int_a^b K(t, s)g(s)K(s, x)ds \\ &= \int_a^b K(s, t)g(s)K(x, s)ds \\ &= K_2(x, t). \end{aligned}$$

For symmetrizable kerns Marty has shown that

1) There exists at least one characteristic constant.

2) All of the characteristic constants are real.

b) Non-homogeneous Boundary Problems.—We next consider the non-homogeneous boundary problem:

$$L(u) + \lambda gu + r = 0, \qquad R_0(u) = 0, \qquad R_1(u) = 0,$$

where r is a given function of x, continuous on $\left[ab\right]$, and $R_0(u)$, $R_1(u)$ are defined by (4).

We are going to show that the boundary problem is equivalent to a non-homogeneous integral equation. To do this we again make use of Hilbert's third fundamental theorem for

$$f = \lambda g u + r, F = u.$$

Since λ, g, r are continuous by hypothesis, this theorem now reads: if $u(x)$ is continuous, then the combined statement u is of class C'', $L(u) + \lambda g u + r = 0$, $R_0(u) = 0$, $R_1(u) = 0$ is equivalent to the statement that

$$u(x) = \int_a^b K(x, t)\Big[\lambda g(t)u(t) + r(t)\Big]dt$$
$$= \lambda \int_a^b K(x, t)g(t)u(t)dt + \int_a^b K(x, t)r(t)dt.$$

Hence we have the theorem:

Theorem X.—*The non-homogeneous boundary problem:*

$$L(u) + \lambda g u + r = 0,\ R_0(u) = 0,\ R_1(u) = 0$$

with the further condition that $u(x)$ is of class C'', is equivalent to the non-homogeneous integral equation

$$u(x) = f(x) + \lambda \int_a^b K(x, t)g(t)u(t)dt$$

where $u(x)$ is continuous and $f(x) = \int_a^b K(x, t)r(t)dt.$

c) The Exceptional Case $\lambda = 0$ Is a Characteristic Constant.—All of the preceding developments have been made under the hypothesis (H_2) that $\lambda = 0$ is not a characteristic constant of the boundary problem. This hypothesis, however, is not satisfied in certain problems of mathematical physics. For example, the following boundary problem in the theory of heat:

$$\frac{d^2u}{dx^2} + \lambda u = 0,\ u'(a) = 0,\ u'(b) = 0$$

has the non-trivial solution u = constant for $\lambda = 0$.

The exceptional case in which the hypothesis (H_2) is not satisfied can be treated, according to Hilbert,[1] by introducing a modified Green's function.

[1] See HILBERT, "Gott. Nach.," p. 213, 1904.

In most cases, however, the following simple artifice, due to Kneser will be sufficient.

Let us replace the assumption (H_2) by the much milder assumption (H_2'):

(H_2') *There exists at least one value c of λ which is not a characteristic constant for the boundary problem.*

Let us write the differential equation

$$\frac{d}{dx}\left(p\frac{du}{dx}\right) + (q + \lambda g)u = 0$$

in the form

$$\frac{d}{dx}\left(p\frac{du}{dx}\right) + \left[q + cg + (\lambda - c)g\right]u = 0$$

or

$$\frac{d}{dx}\left(p\frac{du}{dx}\right) + (\bar{q} + \bar{\lambda}g)u = 0,$$

where

$$\bar{q} = q + cg, \bar{\lambda} = \lambda - c.$$

Then $\bar{\lambda} = 0$ is certainly not a characteristic constant of the boundary problem

$$\frac{d}{dx}\left(p\frac{du}{dx}\right) + (\bar{q} + \bar{\lambda}g)u = 0, R_0(u) = 0, R_1(u) = 0,$$

since, by hypothesis (H_2'), the boundary problem

$$\frac{d}{dx}\left(p\frac{du}{dx}\right) + qu + cgu = 0, R_0(u) = 0, R_1(u) = 0$$

has no solution other than $u \equiv 0$.

II. Applications to Some Problems of the Calculus of Variations

77. Some Auxiliary Theorems of the Calculus of Variations. *a) Formulation of the Simplest Type of Problem.*—For the simplest type of problems of the calculus of variations we have given

1) Two points $P_0(x_0, y_0)$, $P_1(x_1, y_1)$.

2) A function $F(x, y, y')$ of three independent variables.

Required: to find among all curves

$$y = y(x) \tag{30}$$

joining P_0 and P_1 that one which furnishes for the definite integral

$$J(y) = \int_{x_0}^{x_1} F\left[x,\, y(x),\, y'(x)\right] dx, \quad \left[y'(x) = \frac{d}{dx}y(x)\right]$$

the smallest value.

Concerning the admissible curves (30) we make the assumption that they satisfy the following conditions:

$$(31)\quad \begin{aligned} &A)\ y(x) \text{ of class } C''.\\ &B)\ y(x_0) = y_0;\ y(x_1) = y_1. \end{aligned}$$

We assume that the function F is of class C''' for all systems of values x, $y(x)$, $y'(x)$ furnished by all of the admissible curves.

b) Euler's Differential Equation.—Suppose we have found the minimizing curve C_0

$$C_0: \qquad y = f(x).$$

We replace it by a neighboring admissible curve of the special form

$$C: \qquad y = f(x) + \epsilon\eta(x),$$

where ϵ is a small constant and $\eta(x)$ a function of x satisfying the conditions:

$$(32)\quad \begin{aligned} &A')\ \eta(x) \text{ is of class } C''.\\ &B')\ \eta(x_0) = 0,\ \eta(x_1) = 0. \end{aligned}$$

Since C_0 minimizes the integral, we have

$$\int_{x_0}^{x_1} F\left[x, f + \epsilon\eta, f' + \epsilon\eta'\right] dx \geqq \int_{x_0}^{x_1} F\left[x, f, f'\right] dx,$$

which we shall write in the form

$$I(\epsilon) \geqq I(0).$$

Considered as a function of ϵ, $I(\epsilon)$ has, therefore, a minimum for $\epsilon = 0$ and hence

$$I'(0) = 0, \; I''(0) \geqq 0.$$

These are *necessary* conditions for a minimum. It is customary to write

$$\delta I \equiv \epsilon I'(0), \qquad \delta^2 I \equiv \epsilon^2 I''(0), \; (\eta)$$

and to call δI, $\delta^2 I$, the first and second variation respectively.

We first consider the condition $I'(0) = 0$. By definition

$$I(\epsilon) = \int_{x_0}^{x_1} F\left[x, y + \epsilon\eta, y' + \epsilon\eta'\right] dx,$$

whence, by the rules for differentiating a definite integral with respect to a parameter,

$$I'(\epsilon) = \int_{x_0}^{x_1}\left[\bar{F}_y\eta + \bar{F}_{y'}\eta'\right] dx, \; (\eta) \tag{33}$$

where the dash indicates that we use the arguments, x, $f(x) + \epsilon\eta(x)$, $f'(x) + \epsilon\eta'(x)$. Therefore

$$I'(0) = \int_{x_0}^{x_1}\left(F_y\eta + F_{y'}\eta'\right) dx, \; (\eta)$$

the arguments now being x, $f(x)$, $f'(x)$.[1]

An integration by parts gives

$$I'(0) = \Big[\eta F_{y'}\Big]_{x_0}^{x_1} + \int_{x_0}^{x_1} \eta\left(F_y - \frac{d}{dx}F_{y'}\right) dx, \; (\eta)$$
$$= \int_{x_0}^{x_1} \eta\left(F_y - \frac{d}{dx}F_{y'}\right) dx, \; (\eta)$$

on account of (32).

But $I'(0) = 0$, and, hence, by the fundamental theorem of the calculus of variations,[2]

$$F_y - \frac{d}{dx}F_{y'} = 0. \tag{34}$$

[1] Consult Bolza, "Lectures on the Calculus of Variations," pp. 16*ff*., University of Chicago Press, 1904.

[2] See Bolza, "Lectures on the Calculus of Variations," §5.

This is a differential equation of the second order for the determination of $y = f(x)$. It is the first necessary condition and is known as Euler's equation. Its solution involves two arbitrary constants which have to be determined by the two conditions (31).

From (33) we get

$$I''(\epsilon) = \int_{x_0}^{x_1}\left(\bar{F}_{yy}\eta^2 + 2\bar{F}_{yy'}\eta\eta' + \bar{F}_{y'y'}\eta'^2\right)dx,$$

whence

$$I''(0) = \int_{x_0}^{x_1}\left(F_{yy}\eta^2 + 2F_{yy'}\eta\eta' + F_{y'y'}\eta'^2\right)dx, \ (\eta).$$

c) Euler's Rule for Isoperimetric Problems.—For isoperimetric problems the admissible curves are subject to a third condition C in addition to A) and B):

$$C: \qquad \int_{x_0}^{x_1} G\Big[x, y(x), y'(x)\Big] = l, \text{ a constant.}$$

The problem is, then, to determine, among all curves

$$y = y(x)$$

satisfying the conditions A) B) C), that one which will furnish the smallest value for the integral

$$J(y) = \int_{x_0}^{x_1} F(x, y, y')dx.$$

The conditions on G are the same as those on F, *viz.*, F and G are continuous and possess continuous partial derivatives of the first, second, and third order in the region under consideration. Put

$$H = F + \lambda G,$$

λ an arbitrary constant. Then, by Euler's rule, the first necessary condition for a minimum is the same as if it were required to minimize the integral

$$\int_{x_0}^{x_1} H(x, y, y')dx$$

with respect to the totality of curves satisfying the conditions A) and B), that is,

$$H_y - \frac{d}{dx}H_{y'} = 0,$$

or, more explicitly,

$$F_y - \lambda G_y - \frac{d}{dx}(F_{y'} - \lambda G_{y'}) = 0.$$

The solution of this equation involves three arbitrary constants (α, β, λ) which are to be determined by the three conditions B) and C).

78. Dirichlet's Problem.—We now propose to minimize the integral

$$D(u) = \int_a^b \left[p\left(\frac{du}{dx}\right)^2 - qu^2 \right] dx \tag{35}$$

with respect to the totality M of curves satisfying the conditions

$$\begin{cases} u \text{ is of class } C'', \; u(a) = 0, \; u(b) = 0, \\ \displaystyle\int_a^b u^2(x)dx = 1, \end{cases} \tag{36}$$

with the further hypotheses

$$\left.\begin{array}{l} p > 0, \; p \text{ of class } C' \\ q \text{ continuous} \end{array}\right\} \text{ on } \left[ab\right].$$

Hilbert, to whom the following developments are due, calls this problem *Dirichlet's problem*. The problem is an isoperimetrical problem of the type considered in §77, with

$$F = pu'^2 - qu^2, \; G = u^2.$$

Hence

$$H(x, u, u') = pu'^2 - (q + \lambda)u^2$$

and, therefore, Euler's differential equation for H is

$$H_u - \frac{d}{dx}H_{u'} = -2(q + \lambda)u - \frac{d}{dx}(2pu') = 0.$$

Whence

$$\frac{d}{dx}(pu') + (q + \lambda)u = 0,$$

which may be written

$$(37) \qquad L(u) + \lambda u = 0.$$

Every solution of this problem must satisfy this equation and the conditions (36). Now (36) and (37) constitute a boundary problem of the type previously discussed in which

$$g(x) \equiv 1, \left[puu' \right]_a^b = 0,$$

hence the Theorems I to IX hold.

We can, therefore, state at once that this problem (36), (37) has no solutions except when λ is a characteristic constant. We know that

1) The characteristic constants are real.
2) The characteristic constants are infinite in number.
3) For each characteristic constant the index $q = 1$.

Therefore the λ's are distinct, and, since $p > 0$, we have, by Theorem VIII,

$$(38) \qquad \lambda_1 < \lambda_2 < \lambda_3 < \cdots,$$

with a corresponding complete set,

$$\psi_1(x), \psi_2(x), \psi_3(x), \ldots$$

of normalized orthogonal fundamental functions of class C''. We have, then, for $n = 1, 2, \ldots$

$$L(\psi_n) + \lambda_n \psi_n = 0,$$
$$\psi_n(a) = 0, \psi_n(b) = 0,$$
$$(39) \qquad \int_a^b \psi_n^2(x)dx = 1.$$

If, then, Dirichlet's problem has a solution, it must be of the form

$$u = C\psi_n(x), \lambda = \lambda_n.$$

But from (36) we obtain

$$C^2 \int_a^b \psi_n^2(x)dx = 1.$$

Comparing with (39) we find that $C = \pm 1$. The only possible solutions, therefore, are

$$u = \pm \psi_n(x), \lambda = \lambda_n.$$

But in the proof of Theorem VIII we showed that

$$\lambda_n = \int_a^b (p\psi_n'^2 - q\psi_n^2)dx.$$

Therefore

$$\lambda_n = D(\pm \psi_n), \text{ by (35).}$$

Hence, it follows from (38) that

$$D(\pm \psi_1) < D(\pm \psi_n), (n = 2, 3, \ldots).$$

Hence we infer that: *If there exists at all a function which minimizes $D(u)$ with respect to the totality M of all admissible curves, it must be the function*

$$u = \pm \psi_1(x)$$

and λ must have the value λ_1.

b) Sufficiency Proof. α) Transformation of $D(u)$.—By an integration by parts we obtain

$$\int_a^b (pu')u'dx = upu'\Big]_a^b - \int_a^b u\frac{d}{dx}\left(p\frac{du}{dx}\right)dx.$$

But $\qquad upu'\Big]_a^b = 0$, since $u(a) = 0$, $u(b) = 0$.

Therefore

$$D(u) \equiv \int_a^b \left[p\left(\frac{du}{dx}\right)^2 - qu^2\right]dx$$

$$= -\int_a^b u\left[\frac{d}{dx}\left(p\frac{du}{dx}\right) + qu\right]dx.$$

Therefore

$$D(u) = -\int_a^b uL(u)dx.$$

Let us put

$$-L(u) = \omega(x).$$

Then ω is oontinuous and we have

$$D(u) = +\int_a^b u(x)\omega(x)dx.$$

We now apply Hilbert's third fundamental theorem, whence from

$$u \text{ is of class } C'',\ L(u) + \omega = 0,\ u(a) = 0,\ u(b) = 0$$

it follows that

$$u(x) = \int_a^b K(x, t)\omega(t)dt.$$

Multiply both members of this equation by $\omega(x)$ and integrate. We obtain

$$D(u) = \int_a^b \int_a^b K(x, t)\omega(t)\omega(x)dtdx. \tag{40}$$

β) *Applications of the Expansion Theorem.*—From Corollary II to Theorem XVII of §68, we have

$$u(x) = \int_a^b K(x, t)\omega(t)dt = \sum_{\nu=1}^{\infty} \frac{(\omega\psi_\nu)}{\lambda_\nu} \psi_\nu(x),$$

the series being absolutely and uniformly convergent. Let us put

$$C_\nu = (\omega\psi_\nu), \text{ constant,}$$

then

$$u(x) = \sum_{\nu=1}^{\infty} \frac{C_\nu}{\lambda_\nu} \psi_\nu(x). \tag{41}$$

Form

$$u^2(x) = \sum_{\mu,\nu} \frac{C_\mu C_\nu}{\lambda_\mu \lambda_\nu} \psi_\mu(x)\psi_\nu(x).$$

This series is also absolutely and uniformly convergent. We obtain, therefore,

$$\int_a^b u^2(x)dx = \sum_{\mu,\nu} \frac{C_\mu C_\nu}{\lambda_\mu \lambda_\nu} \int_a^b \psi_\mu(x)\psi_\nu(x)dx = 1.$$

But
$$\int_a^b \psi_\mu(x)\psi_\nu(x)dx = \begin{cases} 1, & \mu = \nu \\ 0, & \mu \neq \nu. \end{cases}$$

Therefore

$$\sum_\nu \frac{C_\nu^{\,2}}{\lambda_\nu^{\,2}} = 1. \tag{42}$$

We now apply Corollary III to Theorem XVII of §68, to the double integral in (40). We obtain

$$D(u) = \sum_{\nu=1}^{\infty} \frac{(\omega\psi_\nu)^2}{\lambda_\nu} = \sum_{\nu=1}^{\infty} \frac{C_\nu^{\,2}}{\lambda_\nu}.$$

γ) *Computation of* $D(u) - D(\pm\,\psi_1)$.—Let us now compute the difference

$$D(u) - D(\pm\,\psi_1) = D(u) - \lambda_1.$$

On account of (42)

$$\lambda_1 = \sum_{\nu=1}^{\infty} \frac{C_\nu^{\,2}\lambda_1}{\lambda_\nu^{\,2}}.$$

Therefore

$$D(u) - \lambda_1 = \sum_\nu \frac{C_\nu^{\,2}}{\lambda_\nu} - \sum \frac{C_\nu^{\,2}\lambda_1}{\lambda_\nu^{\,2}}$$
$$= \sum_{\nu=2}^{\infty} \frac{C_\nu^{\,2}}{\lambda_\nu^{\,2}}(\lambda_\nu - \lambda_1) \geq 0,$$

since
$$\frac{C_\nu^{\,2}}{\lambda_\nu^{\,2}} \geq 0, \text{ and } \lambda_\nu - \lambda_1 > 0 \text{ for } \nu \geq 2.$$

Therefore we have

$$D(u) - D(\pm\,\psi_1) \geq 0.$$

The equality holds only if

$$C_\nu = 0 \text{ for } \nu = 2, 3, \ldots, \infty.$$

But in this event from (42) we obtain $C_1 = \pm \lambda_1$, whence from (41), $u(x) = \pm \psi_1(x)$. Therefore, whenever $u(x) \not\equiv \psi_1(x)$, it results that

$$D(u) - D(\pm \psi_1(x)) \geqq 0.$$

Thus we have proved the following theorem:

Theorem XI.—$u = \pm \psi_1(x)$ *furnishes for*

$$D(u) = \int_a^b \left[p\left(\frac{du}{dx}\right)^2 - qu^2 \right] dx$$

a smaller value (viz., λ_1*) than any other function satisfying the conditions*

$$u \text{ is of class } C'', \; u(a) = 0, \; u(b) = 0, \; \int_a^b u^2(x)dx = 1.$$

c) Dropping the Assumption (H_2).—The results for the Dirichlet problem seem to presuppose (H_2), that is, that $\lambda = 0$ is not a characteristic constant of the boundary problem

$$L(u) + \lambda u = 0, \; u(a) = 0, \; u(b) = 0,$$

but they are independent of this assumption.

Proof.—α) We notice first that the assumption, $\lambda = 0$ is not a characteristic constant, was necessary for the construction of a Green's function. But the Green's function did not occur in the proof of Theorem VIII. In this theorem, we made the special assumption that

$$R_0(u) = 0, \; R_1(u) = 0, \text{ imply } \Big[puu' \Big]_a^b = 0.$$

Let then λ_0 be any characteristic constant of the boundary problem:

$$(43) \qquad L(u) + \lambda u = 0, \; R_0(u) = 0, \; R_1(u) = 0$$

and $\psi_0(x)$ a corresponding normalized fundamental function of the boundary problem:

ψ_0 of class C'', $L(\psi_0) + \lambda_0\psi_0 = 0$, $R_0(\psi_0) = 0$, $R_1(\psi_0) = 0$,

$$\int_a^b \psi_0^2(x)dx = 1.$$

Hence we derived, since $\Big[p\psi_0\psi_0'\Big]_a^b = 0$, without using the Green's function, that is, independently of (H_2) the equality

$$\lambda_0 = \int_a^b (p\psi_0'^2 - q\psi_0^2)dx \equiv D(\psi_0.)$$

Hence followed from the corollary to Theorem VIII that

$$p > 0,\ q \leqq 0, \text{ imply } \lambda_0 > 0;$$

that is, all of the characteristic constants of the boundary problem are greater than zero. Thus the hypothesis (H_2) is satisfied and the Theorems V to VIII certainly hold.

β) Suppose now that the condition $q \leqq 0$ is not satisfied. Denote, as before, by m the minimum of $-q$ on $\Big[ab\Big]$, that is,

$$q + m \leqq 0.$$

Then the differential equation

$$L(u) + \lambda u \equiv \frac{d}{dx}\Big(p\frac{du}{dx}\Big) + (q + \lambda)u = 0$$

becomes

$$\bar{L}(u) + \bar{\lambda}u \equiv \frac{d}{dx}\Big(p\frac{du}{dx}\Big) + \bar{q}u + \lambda u = 0$$

where

$$q + m = \bar{q} \text{ and } \lambda - m = \bar{\lambda}.$$

Since $p > 0$, $\bar{q} \leqq 0$, the Theorems V to VIII hold for the boundary problem

(44) $\qquad \bar{L}(u) + \bar{\lambda}u = 0,\ R_0(u) = 0,\ R_1(u) = 0.$

In particular, it has a single infinitude of real, distinct characteristic constants:

$$0 < \bar{\lambda}_1 < \bar{\lambda}_2 < \bar{\lambda}_3 < \ . \ . \ .$$

with corresponding normalized fundamental functions

$$\psi_1(x), \psi_2(x), \psi_3(x), \quad . \quad . \quad .$$

But since

$$\bar{L}(u) = L(u) + mu$$

it follows that, if λ_0 is a characteristic constant and $\psi_0(x)$ a fundamental function for (43), then $\bar{\lambda}_0 = \lambda_0 - m$ is a characteristic constant and $\psi_o(x)$ the corresponding fundamental function for (44), and *vice versa.*

Hence also the boundary problem (43) has an infinitude of real, distinct characteristic constants forming an increasing sequence:

$$\lambda_1 = \bar{\lambda}_1 + m, \lambda_2 = \bar{\lambda}_2 + m, \quad . \quad . \quad ,$$
$$\lambda_1 < \lambda_2 < \lambda_3 < \quad . \quad . \quad . \quad ,$$

with the corresponding normalized orthogonal fundamental functions $\psi_1(x)$, $\psi_2(x)$, ...

γ) Consider now the problem

$$D(u) = \text{minimum on } (M).$$

This problem is evidently equivalent to

$$\bar{D}(u) \equiv D(u) - m = \text{minimum on } (M), \textit{ i.e.},$$

$$\bar{D}(u) = \int_a^b \left[p\left(\frac{du}{dx}\right)^2 - (q + m)u^2 \right] dx = \text{minimum on } (M).$$

Since $q + m \leqq 0$, our former results hold and

$$u = \pm \psi_1(x)$$

furnishes the minimum for $\bar{D}(u)$, that is,

$$\bar{D}\left[\pm \psi_1(x) \right] = \bar{\lambda}_1.$$

On account of the equivalence of the two problems, the same function $u = \pm \psi_1(x)$ furnishes the minimum for $D(u)$ and

$$D(\pm \psi_1) = \bar{D}(\pm \psi_1) + m = \bar{\lambda}_1 + m = \lambda_1.$$

But by β) $\lambda_1 = \bar{\lambda}_1 + m$ is the smallest characteristic constant for the boundary problem

$$L(u) + \lambda u = 0,\ u(a) = 0,\ u(b) = 0.$$

This shows that Theorem XI is true also when the assumption (H_2) is not satisfied.

79. Applications to the Second Variation.—Hilbert has made an application of the preceding result to the discussion of the second variation for the simplest type of variation problem considered in §77.

a) Reduction of the Problem.—The problem is to find the condition under which

$$(45) \qquad \int_{x_0}^{x_1} \left(F_{yy}\eta^2 + 2F_{yy'}\eta\eta' + F_{y'y'}\eta'^2 \right) dx \geqq 0$$

for all functions $\eta(x)$ satisfying the conditions

A') $\eta(x)$ is of class C''.
B') $\eta(x_0) = 0,\ \eta(x_1) = 0.$

The arguments of $F_{yy}, F_{yy'}, F_{y'y'}$ are $x,\ y = f(x),\ y' = f'(x)$ where $f(x)$ is a solution of Euler's differential equation (34) satisfying the initial conditions. We adopt the notation

$$F_{yy}\left[x,\ f(x),\ f'(x)\right] = P(x),\ F_{yy'}\left[x,\ f(x),\ f'(x)\right] = Q(x),$$
$$F_{y'y'}\left[x,\ f(x),\ f'(x)\right] = R(x).$$

Then (45) becomes

$$(46) \qquad \int_{x_0}^{x_1} \left[P\eta^2 + 2Q\eta\eta' + R\eta'^2 \right] dx \geqq 0,\ (\eta).$$

It is easily shown[1] that a first necessary condition for the inequality (46) is that $R \geqq 0$ on $\left[ab\right]$, which is Legendre's condition.

[1] Compare BOLZA, "Lectures on the Calculus of Variations," §11.

We suppose this condition to be satisfied in the stronger form $R > 0$ on $\left[ab\right]$.

We now transform the integral by integrating the second term by parts

$$\int_{x_0}^{x_1} 2Q\eta\eta' dx = \int_{x_0}^{x_1} Q \frac{d}{dx} \eta^2 dx$$
$$= \eta^2\Big]_{x_0}^{x_1} - \int_{x_0}^{x_1} \eta^2 Q' dx$$
$$= -\int_{x_0}^{x_1} \eta^2 Q' dx.$$

Therefore (45) becomes

$$\int_{x}^{x_1} \left[(P - Q')\eta^2 + R\eta'^2 \right] dx \geqq 0, \ (\eta).$$

If for $\qquad \eta, \ x_0, \ x_1, \ R, \ P - Q'$

we put $\qquad u, \ a, \ b, \ p, \quad - q,$

then the above inequality becomes

$$D(u) \geqq 0$$

for the totality N of all functions u satisfying the conditions

$$u \text{ is of class } C'', \ u(a) = 0, \ u(b) = 0.$$

b) *Connection with Dirichlet's Problem.*—We now show the equivalence of the two statements

(47) $\qquad D(u) \geqq 0$, for all curves of class N, and

(48) $\qquad D(u) \geqq 0$, for all curves of class M.

A) The class of curves M is contained within the class N and hence

$$D(u) \geqq 0, \ (N) \text{ implies } D(u) \geqq 0, \ (M).$$

B) Suppose $u \not\equiv 0$ belongs to the class (N). Construct $u_1 = \rho u$ such that

$$\int_a^b u_1^2 \, dx = 1.$$

Then u_1 belongs to M, for u_1 is of class C'' and $u_1(a) = 0$, $u_1(b) = 0$, since $u(a) = 0$, $u(b) = 0$ and by construction

$$\int_a^b u_1^2 dx = 1.$$

Therefore $$D(u_1) \geqq 0.$$

But $$D(u_1) = D(\rho u) = \rho^2 D(u).$$

Therefore $$D(u) \geqq 0.$$

Hence, if u belongs to the class (N) and $u \not\equiv 0$, then $D(u) \geqq 0$, while if $u \equiv 0$ also $D(u) = 0$. Therefore, if u belongs to (N), then $D(u) \geqq 0$.

The equivalence between the two inequalities (47) and (48) being established, we can now apply the results of §78:

The smallest value which $D(u)$ can take in M is λ_1 and this value is furnished by $u = \pm \psi_1(x)$ and by no other function of M.

Hence if

1) $\lambda_1 > 0$, then $D(u) > 0$ (M).

2) $\lambda_1 = 0$, then $D(u) > 0$ (M).

 except when $u = \pm \psi_1(x)$, in which case $D(u) = 0$.

3) $\lambda_1 < 0$, then $D(u)$ can be made negative in M.

Hence, $\lambda_1 \geqq 0$ is the necessary and sufficient condition that $D(u) \geqq 0$ for all curves of the class M and therefore also the necessary and sufficient condition that $D(u) \geqq 0$ for all curves of the class N.

Returning now to the notation of the calculus of variations, we have the theorem:

Theorem XII.—*Suppose* $R > 0$ *and let* λ_1 *denote the smallest characteristic constant of the boundary problem*

$$\frac{d}{dx}\left(R\frac{du}{dx}\right) - (P - Q')u = 0,\ u(x_0) = 0,\ u(x_1) = 0,$$

then $\lambda_1 \geqq 0$ *is the necessary and sufficient condition that*

$$\delta^2 I \geqq 0,\ (\eta).$$

80. Connection with Jacobi's Condition. *a) Sturm's Oscillation Theorem.*—It is *a priori* clear that the condition $\lambda_1 \geqq 0$ must be equivalent to Jacobi's condition.[1] The connection between the two can be established by means of Sturm's oscillation theorem.

Since the differential equation

$$L(u) + \lambda u = 0$$

has no singularities on $\left[ab\right]$, there exists one and only one solution for which

$$u(a) = 0, \; u'(a) = 1.$$

Let us call this $u = V(x, \lambda)$, so that

$$V(a, \lambda) = 0, \; V'(a, \lambda) = 1.$$

Any other solution $u(x)$, for which $u(a) = 0$, is then of the form

$$u = C \cdot V(x, \lambda).$$

Now the boundary problem

$$L(u) + \lambda u = 0, \; u(a) = 0, \; u(b) = 0$$

has for $\lambda = \lambda_1$ a solution $u = \psi_1(x)$, for which

$$\psi_1(a) = 0, \; \psi_1(b) = 0.$$

Hence

$$\psi_1(x) = C \cdot V(x, \lambda_1)$$

and, therefore,

$$V(a, \lambda_1) = 0, \; V(b, \lambda_1) = 0.$$

Let us designate by $\xi(\lambda)$ the root of $V(x, \lambda)$ next greater than a. Then Sturm's oscillation theorem states[2] that

[1] *Cf.* Bolza, "Lectures," §16.

[2] See Böcher, *Bull.* Am. Math. Soc., 4, 1898.

1) as λ increases, $\xi(\lambda)$ decreases

(49) as λ decreases, $\xi(\lambda)$ increases

2) $V(x, \lambda_1) \neq 0$ between a and b and, therefore, $\xi(\lambda_1) = b$.

b) The Conjugate Point.—Consider the solution

$$u = V(x, 0)$$

and put

$$a' = \xi(0).$$

Case I.—$\lambda_1 > 0$. Then, by (49), as λ decreases from λ_1 to 0, $\xi(\lambda)$ increases from $\xi(\lambda_1) = b$ to $\xi(0) = a'$ and, therefore, $a' > b$.

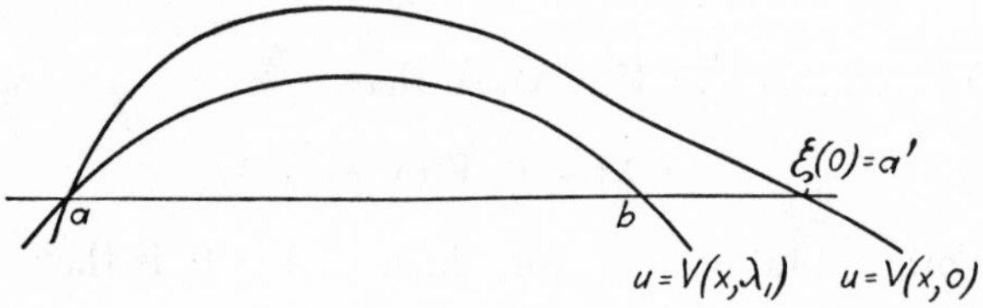

FIG. 19.

Case II.—$\lambda_1 = 0$. Then $a' = b$.

Case III.—$\lambda_1 < 0$. Then, by (49), as λ increases from λ_1 to 0, $\xi(\lambda)$ decreases from $\xi(\lambda_1) = b$ to $\xi(0) = a'$ and, therefore, $a' < b$.

Let us now return to the notation of the calculus of variations, that is

from
$$a,\ b,\ \frac{d}{dx}\left(p\,\frac{du}{dx}\right) + qu$$

to
$$x_0,\ x_1,\ \frac{d}{dx}\left(R\,\frac{du}{dx}\right) - (P - Q')u.$$

Then $u = V(x, 0)$ is defined as that integral of Jacobi's differential equation

$$\frac{d}{dx}\left(R\,\frac{du}{dx}\right) - (P - Q')u = 0$$

which satisfies the initial conditions.

$$u(x_0) = 0, \ u'(x_0) = 1.$$

Hence $V(x, 0)$ is, up to a constant factor, identical with the function denoted in the calculus of variations by $\Delta(x, x_0)$ and, therefore, a' is identical with x_0', the abscissa of the point conjugate to x_0.

This establishes the equivalence between Hilbert's condition

$$\lambda_1 \geqq 0$$

and Jacobi's condition

$$x_1 \leqq x_0'$$

for a non-negative sign of $\delta^2 I$.

III. Vibration Problems

81. Vibrating String. *a) Reduction to Boundary Problem.* For the homogeneous string we had

$$\frac{\partial^2 \eta}{\partial t^2} = C^2 \frac{\partial^2 \eta}{\partial x^2}. \tag{50}$$

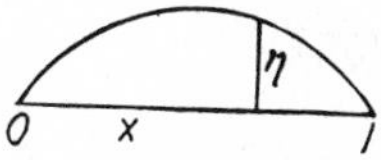

FIG. 20.

$$\eta(0, t) = 0, \ \eta(1, t) = 0, \ (t) \tag{51}$$

$$\eta(x, 0) = f(x), \ \eta_t(x, 0) = F(x), \ (x) \tag{52}$$

where

$$f(0) = 0, \ f(1) = 0$$
$$F(0) = 0, \ F(1) = 0.$$

C was a real positive constant and

$$C^2 = \frac{P}{\kappa\sigma}, \text{ where}$$

P = normal tension, κ = density, σ = area of cross-section.

The differential equation (50) also holds[1] when κ, σ are given functions of x, *i.e.*, the string is non-homogeneous. C^2 is then an always positive, given function of x.

[1] See Weber, "Differentialgleichungen," vol. 2, p. 201.

We try for a solution of the form

$$\eta = u(x) \cdot \phi(t).$$

Substitution of this expression for η in (50) gives

$$u(x)\phi''(t) = C^2 u''(x)\phi(t)$$

whence

$$\frac{\phi''(t)}{\phi(t)} = \frac{C^2 u''(x)}{u(x)}.$$

But the left-hand member is a function of t alone, the right-hand member is a function of x alone; they are equal, hence equal to the same constant, say $-\lambda$. We are thus led to two ordinary differential equations:

$$\frac{d^2u}{dx^2} + \frac{\lambda}{C^2}u = 0 \tag{53}$$

with the boundary conditions

$$u(0) = 0, \; u(1) = 0, \tag{54}$$

and

$$\frac{d^2\phi}{dt^2} + \lambda\phi = 0. \tag{55}$$

b) The Boundary Problem.—The problem (53) (54) is of the type

$$\begin{cases} L(u) + \lambda g u = 0 \\ R_0(u) = 0, \; R_1(u) = 0, \text{ with } \Big[puu' \Big]_a^b = 0 \end{cases} \tag{56}$$

with $p \equiv 1$, $q \equiv 0$, $g = \dfrac{1}{C^2} > 0$ on $\Big[ab \Big]$.

We shall show first that every characteristic constant of (56) is positive if

$$p > 0, \; q \leqq 0, \; g > 0. \tag{57}$$

Let λ_0 be a characteristic constant and $\varphi_0(x)$ a corresponding fundamental function of (56):

$$\varphi_0(x) \text{ of class } C'', \; \varphi_0(x) \not\equiv 0,$$
$$L(\varphi_0) + \lambda_0 g \varphi_0 = 0, \; R_0(\varphi_0) = 0, \; R_1(\varphi_0) = 0.$$

Then

$$\lambda_0 \int_a^b g\varphi_0^2 dx = -\int_a^b \varphi_0 L(\varphi_0) dx$$
$$= \int_a^b \left[p\left(\frac{d\varphi_0}{dx}\right)^2 - q\varphi_0^2 \right] dx.$$

Whence, on account of (57),

$$\lambda_0 = \frac{\int_a^b \left[p\left(\frac{d\varphi_0}{dx}\right)^2 - q\varphi_0^2 \right] dx}{\int_a^b g\varphi_0^2 dx} > 0.$$

Hence, under the assumption (57) the condition (H_2) is always satisfied.

Applying this result to the special case (53) (54), we obtain the following lemma:

Lemma I.—For the boundary problem (53) (54), *all of the characteristic constants are positive.*

Green's function $K(x, t)$ is the same as for the homogeneous string, since it depends only on $L(u)$ and not on g. Therefore

$$K(x, t) = \begin{cases} (1 - t)x, & x \leq t \\ (1 - x)t, & x \geq t. \end{cases}$$

The boundary problem (53) (54) is equivalent to

$$u(x) = \lambda \int_0^1 \frac{K(x, t)}{C^2(t)} u(t) dt.$$

In this integral equation the kernel $\dfrac{K(x, t)}{C^2(t)}$ is not symmetric. But, since $C(x) \neq 0$, this integral equation is reducible to one in which the kernel is symmetric by putting

(58) $u(x) = C(x)\bar{u}(x)$ and $K(x, t) = C(x)C(t)\bar{K}(x, t)$.

Then

$$\bar{u}(x) = \lambda \int_0^1 \bar{K}(x, t)\bar{u}(t) dt.$$

By the same transformation (58) the differential equation (53) is transformed into

$$\frac{d}{dx}\left(C^2\frac{d\bar{u}}{dx}\right) + CC''\bar{u} + \lambda\bar{u} = 0 \tag{59}$$

with the boundary conditions

$$\bar{u}(0) = 0, \ \bar{u}(1) = 0 \tag{60}$$

and it is easily shown that $\bar{K}(x, t)$ is Green's function belonging to the boundary problem (59) (60).

We can now apply the Theorems *V* to *VIII*, with the following result:

The boundary problem (59) (60) has an infinitude of real characteristic constants all of index 1, forming an increasing sequence

$$\lambda_1 < \lambda_2 < \lambda_3 < \ . \ . \ .$$

with corresponding normalized fundamental functions

$$\bar{\psi}_1(x), \bar{\psi}_2(x), \bar{\psi}_3(x), \ . \ . \ . \ .$$

But, if for a given λ, the boundary problem (59) (60) has a non-trivial solution, then for the same λ, (53) (54) has a non-trivial solution $u = C\bar{u}$. Therefore we have the following lemma:

Lemma II.—The boundary problem (53) (54) *has an infinitude of real positive characteristic constants, all of index* 1, *forming an increasing sequence:*

$$0 < \lambda_1 < \lambda_2 < \ . \ . \ . \ ,$$

with corresponding normalized fundamental functions

$$\psi_1(x), \psi_2(x), \ . \ . \ .$$

where

$$\psi_n(x) = C\bar{\psi}_n(x).$$

c) The Generalized Fourier Series.—We return now to (55) with $\lambda = \lambda_n$:

$$\frac{d^2\phi}{dt^2} + \lambda_n\phi = 0. \tag{61}$$

Since $\lambda_n > 0$, the general solution of (61) is

$$\phi = A_n \cos \sqrt{\lambda_n}t + B_n \sin \sqrt{\lambda_n}t.$$

Therefore, presumably a solution of (50) which satisfies (51) is

$$\eta = \sum_{n=1}^{\infty} (A_n \cos \sqrt{\lambda_n}t + B_n \sin \sqrt{\lambda_n}t)\psi_n(x). \tag{62}$$

In order to be a solution, the condition (52) must be satisfied. Imposing condition (52), we obtain

$$\begin{cases} \eta(x, 0) = \sum_{n=1}^{\infty} A_n\psi_n(x) = f(x) \\ \eta_t(x, 0) = \sum_{n=1}^{\infty} \sqrt{\lambda_n}B_n\psi_n(x) = F(x). \end{cases} \tag{63}$$

But $f(x)$ and $F(x)$ are given functions. Hence (63) can be satisfied if A_n and B_n can be determined so that the series in (63) represents $f(x)$ and $F(x)$. Since $f(0) = 0$, $f(1) = 0$, $F(0) = 0$, $F(1) = 0$, it follows, from Theorem VI, that $f(x)$ and $F(x)$ can be so represented if f and F are of class C'', and then

$$A_n = (f\psi_n), \sqrt{\lambda_n}B_n = (F\psi_n).$$

We can be sure that (62) satisfies the differential equation if the series is twice differentiable, term by term, with respect to x and t. Since

$$\psi_n''(x) = -\frac{\lambda_n}{C^2}\psi_n(x)$$

this means if

$$\sum \lambda_n(A_n \cos \sqrt{\lambda_n}t + B_n \sin \sqrt{\lambda_n}t)\psi_n(x)$$

is uniformly convergent in x and t.

d) Special Case of Homogeneous String.—If c is constant, then

$$\lambda_n = n^2\pi^2c^2$$

and

$$\psi_n(x) = \sqrt{2} \sin n\pi x,$$

while

$$\eta = \sum_{n=1}^{\infty} (A_n \cos n\pi ct + B_n \sin n\pi ct) \sin n\pi x$$

$$\equiv \sum_{n=1}^{\infty} \eta_n, \text{ say.}$$

For $f(x)$ and $F(x)$ we obtain:

$$f(x) = \sum_{n=1}^{\infty} \sqrt{2} A_n \sin n\pi x$$

$$F(x) = \sum_{n=1}^{\infty} \sqrt{2} n\pi c\, B_n \sin n\pi x.$$

These are sine series for $f(x)$ and $F(x)$. For the development of an arbitrary function in trigonometric series we need know only that the function is continuous and has a finite number of maxima and minima. These conditions are not so strong as those obtained by means of the theory of integral equations which were demanded for the development in series of fundamental functions.

η_1 is periodic in t with period $T_1 = \frac{2}{c}$, this being the period of the fundamental tone. η_n is periodic with period $T_n = \frac{2}{nc} = \frac{T_1}{n}$, and with intensity $\sqrt{A_n^2 + B_n^2}$. Upon the intensity of the different harmonics depends the quality of the tone. The tone of period $\frac{T}{n}$ is called the nth harmonic overtone, or simply the nth harmonic.

For the non-homogeneous string η_n is also periodic with period $\frac{2\pi}{\sqrt{\lambda_n}}$.

T_n decreases with n; the different periods are not fractions of T_1, hence the total motion is not periodic.

82. Vibrations of a Rope. *a) Differential Equation of the Problem.*—Let us consider a heavy rope of length 1

$$AB = 1$$

suspended at one end A. It is given a small initial displacement in a vertical plane through AB and then each particle is given an initial velocity. The rope is supposed to vibrate in a given vertical plane and the displacement is so small that each particle is supposed to move horizontally; the cross-section is constant; the density constant; the cross-section infinitesimal compared to the length.

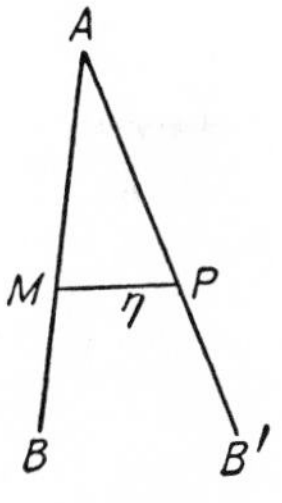

FIG. 21.

Let AB' be the position of the rope at time t and P any point on AB'. Draw PM horizontal and put

$$MP = \eta, \ BM = x.$$

Then the differential equation[1] of the motion is given by

$$\frac{\partial^2\eta}{\partial t^2} = C^2\frac{\partial}{\partial x}\left(x\frac{\partial\eta}{\partial x}\right) \tag{64}$$

where $$C^2 = \text{constant},$$

with the boundary conditions

$$\eta(1, t) = 0, \qquad \eta(0, t) \text{ finite} \tag{65}$$
$$\eta(x, 0) = f(x), \quad \eta_t(x, 0) = F(x). \tag{66}$$

We try for a solution of the form

$$\eta = u(x)\cdot\phi(t).$$

Substitute this expression for η in (64). We obtain

$$u(x)\phi''(t) = C^2\phi(t)\frac{d}{dx}\left(x\frac{du}{dx}\right).$$

[1] See KNESER, "Die Integralgleichungen und ihre Anwendung in der Mathematischen Physick," §11.

Whence

$$\frac{\phi''(t)}{\phi(t)} = C^2 \frac{\dfrac{d}{dx}\left(x\dfrac{du}{dx}\right)}{u(x)} = -\lambda C^2, \text{ constant.}$$

That is

$$\frac{d^2\phi}{dt^2} + \lambda C^2 \phi = 0,$$

and

$$\frac{d}{dx}\left(x\frac{du}{dx}\right) + \lambda u = 0 \tag{67}$$

with the boundary condition derived from (65):

$$u(0) \text{ finite}, \ u(1) = 0. \tag{68}$$

Equation (67) is of the form

$$L(u) + \lambda u = 0$$

with

$$p = x, \ q = 0.$$

In the general case $p \neq 0$ on $\Big[ab\Big]$. This condition is not fulfilled in (67), for $p(0) = 0$. We have also the condition $u(0)$ finite, which did not appear in the general case. The differential equation (67) has a singular point for $x = 0$.

b) Solution of the Boundary Problem.—We solve the boundary problem (67), (68) directly. Put

$$x = \frac{t^2}{4\lambda},$$

then (67) becomes

$$t\frac{d^2u}{dt^2} + \frac{du}{dt} + tu = 0. \tag{69}$$

We try to find a solution of the form

$$u = \sum_{n=0}^{\infty} C_n t^n.$$

If we substitute this series for u in (69), we obtain

$$C_1 + \sum_{n=1}^{\infty}\Big[C_{n-1} + (n+1)^2 C_{n+1}\Big] t^n \underset{t}{\equiv} 0.$$

Therefore

$$C_1 = 0,\ C_{n-1} + (n+1)^2 C_{n+1} = 0,\ n = 1, 2, \dots.$$

That is,

$$C_{n+1} = -\frac{C_{n-1}}{(n+1)^2}.$$

Whence

$$C_1 = C_3 = C_5 = \dots = C_{2\nu+1} = 0 \dots$$

$$C_2 = -\frac{C_0}{2^2},\ C_4 = \frac{C_0}{2^2.4^2},\ C_6 = -\frac{C_0}{2^2.4^2.6^2}, \dots.$$

Therefore

$$u = C_0\left[1 - \frac{t^2}{2^2} + \frac{t^4}{2^2.4^2} - \frac{t^6}{2^2.4^2.6^2} + \dots\right].$$

By comparison with the series for e^x, which is permanently convergent, we show that this series for u is permanently convergent. This series also satisfies (69). Put

$$(70) \qquad J(t) = 1 - \frac{t^2}{2^2} + \frac{t^4}{2^2.4^2} - \frac{t^6}{2^2.4^2.6^2} + \dots$$

$J(t)$ is called Bessel's function of order zero.

$J(t)$ is then a solution of (69). Knowing a particular solution, we can, by means of Green's formula, find the general solution:

$$t\left[J(t)V' - VJ'(t)\right] = C_1 \neq 0 (C_1 \text{ constant}).$$

Divide both members by $J^2(t)$ and integrate. We find

$$\frac{V}{J(t)} = C_1 \int \frac{dt}{tJ^2(t)} + C_0.$$

But from (70)

$$\frac{1}{J^2(t)} = 1 + t^2 P(t^2),$$

$P(t^2)$ being a power series in t^2, whence

$$\int \frac{dt}{tJ^2(t)} = \log t + P_1(t^2)$$

and

$$V = C_0 J(t) + C_1 J(t) \left\{\log t + P_1(t^2)\right\}$$

where $P_1(t^2)$ is a power series in t^2.

This is the general solution of (69). The solution of (67) is given by putting

$$t = 2\sqrt{\lambda}\sqrt{x}.$$

Whence

$$u = C_0 J(2\sqrt{\lambda x}) + C_1 J(2\sqrt{\lambda x})\left\{\log 2\sqrt{\lambda x} + P_1(4\lambda x)\right\}.$$

But, $u(0)$ finite, is a condition upon the solution and this is impossible unless $C_1 = 0$. Therefore

(71) $$u = C_0 J(2\sqrt{\lambda x})$$

is the most general solution of (67), which satisfies the first initial condition. We have the further condition $u(1) = 0$, whence

$$J(2\sqrt{\lambda}) = 0.$$

The solution of this equation gives us the characteristic constants.

c) *Construction of Green's Function.*—We construct for the boundary problem (67) (68) the Green's function $K(x, t)$ satisfying the following conditions:

A) K continuous on $\left[01\right]$.

B) K of class C'' on $\left[0t\right]\left[t1\right]$ separately.

$$\frac{d}{dx}\left(x\frac{dK}{dx}\right) = 0 \text{ on } \left[0t\right]\left[t1\right] \text{ separately.}$$

C) $K(0, t)$ finite, $K(1) = 0$.

D) $K'(t - 0) - K'(t + 0) = \dfrac{1}{t}$.

Integrating the differential equation in *B*), we obtain

$$K(x, t) = \begin{cases} \alpha_0 \log x + \beta_0, & \left[0t\right] \\ \alpha_1 \log x + \beta_1, & \left[t1\right]. \end{cases}$$

But $\quad K(0, t)$ is finite, therefore $\alpha_0 = 0$,
and $\quad K(1, t) = 0, \quad$ therefore $\beta_1 = 0$.
Hence

$$K(x, t) = \begin{cases} K_0(x, t) = \beta_0 \qquad , & [0t] \\ K_1(x, t) = \alpha_1 \log x, & [t1]. \end{cases}$$

From condition A)

$$\beta_0 = \alpha_1 \log t.$$

From condition D), since $K'(t - 0) = 0$ and $K'(t + 0) = \dfrac{\alpha_1}{t}$, we obtain $\alpha_1 = -1$.
Therefore

$$(72) \qquad K(x, t) = \begin{cases} K_0(x, t) = -\log t, & [0t] \\ K_1(x, t) = -\log x, & [t1]. \end{cases}$$

We observe that $K(x, t)$ is symmetric. The graph of $K(x, t)$ for t fixed is shown by the full line in the accompanying figure. As a function of the two variables x and t, $K(x, t)$ is continuous on $[01]$ except at $x = t = 0$. For in the region marked I we have $t > x$ and

$$K(x, t) = \log \frac{1}{t},$$

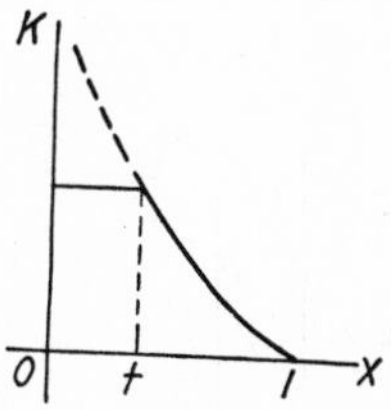

Fig. 22.

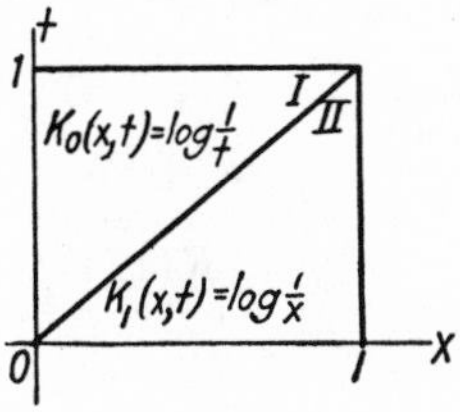

Fig. 23.

while in the region marked II we have $t < x$ and

$$K(x, t) = \log \frac{1}{x}.$$

For $x = 0$, $t > 0$, we have $K = -\log t$, finite.
For $t = 0$, $x > 0$, we have $K = -\log x$, finite.

d) Equivalence with a Homogeneous Integral Equation.—We have

$$L(u) = -\lambda u$$
$$L(K) = 0.$$

Multiply the first of these by $-K$ and the second by u and add. We obtain

$$uL(K) - KL(u) = \lambda uK.$$

Integrate both members of this expression from 0 to $t - 0$ and from $t + 0$ to 1 with respect to t and add. We obtain

$$u(x) = \lambda \int_0^1 K(x, t)u(t)dt.$$

The details of the integration are the same as those given several times previously, except for the term

$$x\Big[uK' - Ku'\Big] \text{ for } x = 0.$$

But here u and K are finite by hypothesis. From (71) we see that if u is finite so also is u'. From the explicit expression (72) for K we find $K'(0, t) \equiv 0$. Therefore

$$x\Big[uK' - K'u\Big] = 0 \text{ for } x = 0.$$

The kernel

$$K(x, t) = \begin{cases} -\log t, & \Big[0t\Big] \\ -\log x, & \Big[t1\Big] \end{cases}$$

is symmetric, but it is discontinuous at one point, *viz.*, $x = t = 0$. Schmidt (page 21 of his dissertation) has shown that the results of the Hilbert-Schmidt theory of continuous symmetric kernels still hold for a discontinuous kernel if

1) $\int_a^b K(x, t)f(t)dt$ for f continuous, is continuous in x on $\left[ab\right]$.

2) The iterated kernel $K_2(x, t)$ is continuous and does not vanish identically.

These conditions are satisfied in the present instance and thus all of the Hilbert-Schmidt theory, as well as the Theorems V to VIII on boundary problems, remain true. Therefore we know that there exists an infinite sequence of positive characteristic constants

$$0 < \lambda_1 < \lambda_2 < \lambda_3 < \ . \ . \ . \ ,$$

with a corresponding complete normalized orthogonal system of fundamental functions. Therefore

$$J(2\sqrt{\lambda}) = 0$$

has an infinitude of positive roots λ_n. Put

$$2\sqrt{\lambda} = k.$$

Then $J(k) = 0$, and the roots are

$$k_n = 2\sqrt{\lambda_n}.$$

The first four values of k_n are[1]

$$k_1 = 2.405,\ k_2 = 5.520,\ k_3 = 8.654,\ k_4 = 11.792$$

and generally $(n - \frac{1}{2})\pi < k_n < n\pi$. Therefore

$$\varphi_n(x) = C_0 J(2\sqrt{\lambda_n x}) = C_0 J(k_n\sqrt{x}).$$

These fundamental functions $\varphi_n(x)$ will become orthogonalized if we choose

$$C_0 = \frac{1}{\sqrt{\int_0^1 J^2(k_n\sqrt{x})dx}}$$

[1] See Fricke, "Analytische-Funktionentheoretische Vorlesungen," p. 74.

But[1]
$$\int_0^1 J^2(k_n\sqrt{x})dx = \frac{1}{[J'(k_n)]^2}.$$

Therefore

$$\psi_n(x) = \frac{J(k_n\sqrt{x})}{J'(k_n)}, \text{ and}$$

$$\eta = \sum_{n=1}^{\infty}\left(A_n \cos\frac{Ck_n}{2}t + B_n \sin\frac{Ck_n}{2}t\right)\psi_n(x).$$

This expression for η satisfies (64) and (65). We now determine A_n, B_n, if possible, in order that (66) may be satisfied. This gives us the two equations

$$\sum A_n\psi_n(x) = f(x)$$
$$\sum \frac{Ck_n}{2}B_n\psi_n(x) = F(x).$$

Since
$$f(0) \text{ is finite, } f(1) = 0,$$
$$F(0) \text{ is finite, } F(1) = 0,$$

$f(x)$ and $F(x)$ can be expanded as series in $\psi_n(x)$, provided f and F are of class C'', and then

$$A_n = \int_0^1 f(x)\psi_n(x)dx = \int_0^1 f(x)\frac{J(k_n\sqrt{x})}{J'(k_n)}dx,$$

while
$$\frac{Ck_n}{2}B_n = \int_0^1 F(x)\psi_n(x)dx.$$

83. The Rotating Rope.—(See KNESER, page 46.)

a) The Problem and Its Differential Equation.—Let FG be an axis around which a plane is rotating with constant velocity; a rope AB is attached at a point A of the axis and constrained to remain in the rotating plane. The velocity of the rotation is so large that the weight of the rope can be neglected. Then the straight line AB perpendicular to FG

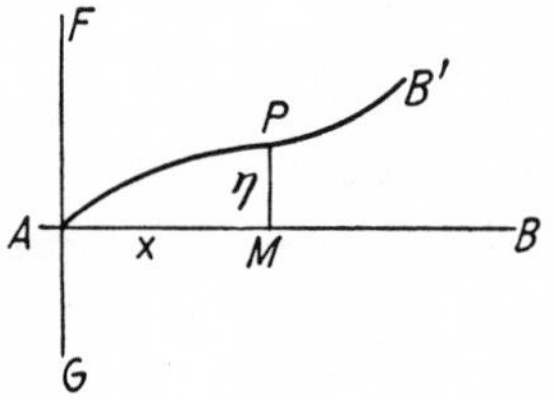

FIG. 24.

[1] See FRICKE, *Loc. cit.*, p. 65.

is a relative position of equilibrium for the rope. Displace the rope slightly from this position AB, then let it go after imparting to its particles initial velocities perpendicular to AB. The rope will then describe small vibrations around the position of equilibrium.

Let APB' be the position of the rope at the time t, P one of its points, $PM \perp AB$. Put $AM = x$, $MP = \eta$, and suppose $AB = 1$. Then the function $\eta(x, t)$ must satisfy the partial differential equation

$$\frac{\partial^2 \eta}{\partial t^2} = C^2 \frac{\partial}{\partial x}\left[(1 - x^2)\frac{\partial \eta}{\partial x}\right], \quad C \text{ constant},$$

the boundary conditions

$$\eta(0, t) = 0, \ \eta(1, t) = \text{finite}, \ (t),$$

and the initial conditions

$$\eta(x, 0) = f(x), \ \eta_t(x, 0) = F(x), \ (x).$$

Putting
$$\eta = u(x)\,\phi\,(t)$$

we obtain

$$\frac{d}{dx}\left[(1 - x^2)\frac{du}{dx}\right] + \lambda u = 0 \tag{73}$$

$$u(0) = 0, \ u(1) = \text{finite}$$

$$\frac{d^2\phi}{dt^2} + \lambda C^2 u = 0.$$

b) Solution of the Boundary Problem.—$x = 1$ is a singular point of (73). By Fuch's theory[1] (73) has for every λ one and only one solution, which in the vicinity of $x = 1$ is given by

$$u_1 = 1 + a_1(x - 1) + a_2(x - 1)^2 + \cdots$$

Every other solution u_2 is obtained from Green's formula

$$(1 - x^2)(u_1u_2' - u_2u_1') = C_1 \neq 0.$$

[1] See Goursat, "Cours D'Analyse," tome 2, §412.

Therefore

$$\frac{d}{dx}\frac{u_2}{u_1} = \frac{C_1}{(x^2 - 1)u_1{}^2} = C_1\left\{\frac{1}{x-1} + S_1(x-1)\right\}.$$

Whence

$$u_2 = C_0u_1 + C_1u_1\left\{\log(x-1) + S(x-1)\right\}.$$

Therefore the condition, $u(1)$ finite, leads to $C_1 = 0$ and hence

(74) $$u = C_0u_1$$

is the most general solution of (73) which satisfies the second boundary condition. Put

$$u_1 = U(x, \lambda).$$

Then the first boundary condition gives

$$U(0, \lambda) = 0$$

for the determination of the characteristic constants. From (74) we see that each characteristic constant is of index 1.

Let λ_0 be a characteristic constant and $\varphi_0(x) = U(x, \lambda_0)$ a corresponding fundamental function. Then $\varphi_0(x)$ is an odd function:

$$\varphi_0(-x) = -\varphi(x).$$

For u_1 can be expanded according to powers of x, say

$$u_1 = \sum_{r=1}^{\infty} C_r x^r.$$

The substitution of this series in the differential equation leads to the following recurrent formula

$$C_{r+2} = \frac{r(r+1) - \lambda_0}{(r+1)(r+2)} C_r,\ r = 0, 1, 2, \ldots,$$

which shows that if $C_0 = 0$, then

$$C_{2\mu} = 0,\ \mu = 0, 1, 2, \ldots.$$

But from $U(0, \lambda) = 0$, we obtain $\varphi_0(0) = 0$, hence $C_0 = 0$, and, therefore, $\varphi_0(x)$ is an odd function.

We can at once indicate some of the characteristic constants and fundamental functions from the theory of Legendre's polynomials:

$$P_n(x) = \frac{1}{2^n n!}\frac{d^n}{dx^n}\left[(x^2 - 1)^n\right]$$
$$P_0(x) = 1.$$

$P_n(x)$ is a rational integral function of degree n, satisfying the differential equation

$$\frac{d}{dx}\left[(1 - x^2)\frac{dy}{dx}\right] + n(n + 1)y = 0.$$

Furthermore, $P_{2n}(x)$ is an even function and $P_{2n}(0) \neq 0$, $P_{2n-1}(x)$ is an odd function and $P_{2n-1}(0) = 0$. Therefore

$$\lambda_n = 2n(2n - 1),\ n = 1, 2, 3, \ldots$$

is a characteristic constant and $P_{2n-1}(x)$ a corresponding fundamental function.

The characteristic constants are of index 1 since the condition $u(1) = 0$ determines u up to a constant factor.

We will now show that there are no other characteristic constants than

$$\lambda_n = 2n(2n - 1),\ n = 1, 2, 3, \ldots$$

Suppose λ_0 to be a characteristic constant and $\lambda_0 \neq \lambda_n$ and $\psi_0(x)$ a corresponding normalized fundamental function, then, according to the orthogonality theorem (§59),

$$\int_0^1 \psi_0(x)P_{2n-1}(x)dx = 0.$$

Now $\psi_0(x)$ is continuous on the interval $\left[-1, +1\right]$ and, therefore, by a theorem due to Weierstrass,[1] can be expanded in an infinite series of polynomials uniformly convergent on $\left[-1, +1\right]$:

[1] GOURSAT-HEDRICK, "Mathematical Analysis," vol. 1, §199.

(75) $$\psi_0(x) = \sum_{r=1}^{\infty} G_r(x).$$

But, as shown under *b*),

$$\psi_0(-x) = -\psi_0(x).$$

Therefore, from

$$\psi_0(-x) = \sum_{r=1}^{\infty} G_r(-x)$$

and (75) above, we obtain

$$\psi_0(x) = \frac{1}{2}\sum_{r=1}^{\infty}\Big[G_r(x) - G_r(-x)\Big] \equiv \sum_{r=1}^{\infty} H_r(x)$$

and $\sum H_r(x)$ is uniformly convergent on $\Big[-1, +1\Big]$. From the definition of $H_r(x)$ we obtain

$$H_r(-x) = -H_r(x),$$

that is, $H_r(x)$ is an odd function.

From the uniform convergence of $\sum H_r(x)$ we have that for every positive ϵ it is possible to find an m such that

(76) $$\left|\psi_0(x) - \sum_{r=1}^{m} H_r(x)\right| < \epsilon, \quad \Big[-1, +1\Big].$$

Let us choose

(77) $$0 < \epsilon < \frac{1}{\int_0^1 |\psi_0(x)|dx}.$$

Since $H_r(x)$ is an odd function we have

(78) $$\sum_{r=1}^{m} H_r(x) = \sum_{r=0}^{n} C_r x^{2r-1}.$$

But $P_{2n-1}(x)$ are odd functions:

$$P_1(x) = a_{11}x$$
$$P_3(x) = a_{22}x^3 + a_{21}x$$
$$\cdots\cdots\cdots\cdots$$
$$P_{2n-1}(x) = a_{nn}x^{2n-1} + \cdots + a_{n1}x$$

with $$a_{11}, a_{22}, \ldots, a_{nn} \neq 0.$$

These equations can be solved sequentially for x, x^3, $\ldots$, x^{2n-1} in terms of P_1, P_3, $\ldots$, P_{2n-1}. Put these values of x^{2r-1} in (78) above and we will have

$$\sum_{r=1}^{m} H_r(x) = \sum_{r=0}^{n} C_r P_{2r-1}(x).$$

This expression for $\sum H_r(x)$, substituted in (76), gives

$$\left| \psi_0(x) - \sum_{r=0}^{n} C_r P_{2r-1}(x) \right| < \epsilon.$$

Put

(79) $$\psi_0(x) = \sum_{r=0}^{n} C_r P_{2r-1}(x) + r(x),$$

whence

$$|r(x)| < \epsilon \text{ on } \Big[-1, +1 \Big].$$

From (79) we obtain

$$\int_0^1 \psi_0^2(x)dx = \sum_{r=0}^{n} C_r \int_0^1 \psi_0(x)P_{2r-1}(x)dx + \int_0^1 r(x)\psi_0(x)dx.$$

But $$\int_0^1 \psi_0(x)P_{2r-1}(x)dx = 0$$

and $$\int_0^1 \psi_0^2(x)dx = 1.$$

Therefore $$1 = \int_0^1 r(x)\psi_0(x)dx.$$

Now $\left|\int_0^1 r(x)\psi_0(x)dx\right| \leq \int_0^1 \left|r(x)\psi_0(x)\right| dx$

$$< \epsilon \int_0^1 \left|\psi_0(x)\right| dx.$$

Therefore
$$1 < \epsilon \int_0^1 \left|\psi_0(x)\right| dx,$$

which, on account of our choice (77) of ϵ, gives

$$1 < \epsilon \int_0^1 \left|\psi_0(x)\right| dx < 1,$$

which constitutes a contradiction. Therefore

$$\lambda_n = 2n\ (2n - 1)$$

are the only characteristic constants, and the only fundamental functions are

$$\varphi_n(x) = \frac{P_{2n-1}(x)}{\sqrt{\int_0^1 P^2{}_{2n-1}(x)dx}}.$$

But
$$\int_0^1 P^2{}_{2n-1}(x)dx = \frac{1}{4n-1},$$

as shown in the theory of Legendre's polynomials. Therefore

$$\varphi_n(x) = \sqrt{4n - 1}\, P_{2n-1}(x).$$

C) Equivalence with Integral Equation.—We construct the Green's function as before and obtain

$$K(x, t) = \begin{cases} \frac{1}{2} \log \frac{1+x}{1-x}, & 0 \leq x \leq t \\ \frac{1}{2} \log \frac{1+t}{1-t}, & t \leq x \leq 1. \end{cases}$$

$K(x, t)$ is symmetric and has but one point of discontinuity and that is for $x = t = 1$.

Schmidt's conditions for a discontinuous kernel hold, however, and so the theorems of the Hilbert-Schmidt theory apply. Proceeding as in the previous problems, we find that the boundary problem is equivalent to the following integral equation:

$$u(x) = \lambda \int_0^1 K(x, t)u(t)dt.$$

The Theorems V to IX hold for this boundary problem as well as the orthogonality theorem, §59. Hence

$$\int_0^1 P_{2n-1}(x)P_{2m-1}(x)dx = 0, \; m \neq n$$

This last result agrees with a result of the theory of Legendre polynomials. Applying Theorem VI, we see that if f is of class C'' and $f(0) = 0$, then

$$f(x) = \sum_{r=0}^{\infty} C_r P_{2n-1}(x)$$

where $$C_r = (fP_{2n-1})\frac{4n-1}{1}.$$

IV. Applications of the Hilbert-Schmidt Theory to the Flow of Heat in a Bar

84. The Partial Differential Equations of the Problem.[1] *a) General Hypotheses.*—The theory of the flow of heat in a bar is based upon the following hypotheses:

A) Let dm denote the mass of an element of a conductor of temperature θ, C its specific heat, then the amount of heat dQ necessary to increase the temperature of the element from θ to $\theta + d\theta$ is given by the formula

$$dQ = Cdmd\theta.$$

B) Let $d\omega$ be the area of an element of surface through an interior point P of a conductor, n one of the two normals to the element, k the inner conductivity at point P, then the amount of heat which flows through the inner surface $d\omega$ in time dt is

$$dQ = -k\frac{\partial\theta}{\partial n}d\omega dt.$$

[1] See Weber, "Partielle Differentialgleichungen," §§32–34.

C) Let $d\omega$ be an element of the outer surface of the conductor, h the outer conductivity at P, θ the temperature of the conductor at P, Θ the temperature of the surrounding medium at P, then the amount of heat which flows through the outer surface $d\omega$ in time dt is

$$dQ = h(\theta - \Theta)d\omega dt.$$

b) We now apply these principles to the determination of the flow of heat in a straight bar placed along the x-axis of a rectangular system of coordinates, with a cross-section σ, infinitesimal compared to the length, so that the temperature θ may be considered as constant; accordingly, θ will be a function of x and t: $\theta(x, t)$. The bar is imbedded in a medium of given temperature which is also supposed to be a function of x and t: $\Theta(x, t)$. The bar has a given initial temperature $\theta(x, 0) = f(x)$. Consider an element of the bar of length dx; we may regard it as a cylinder.

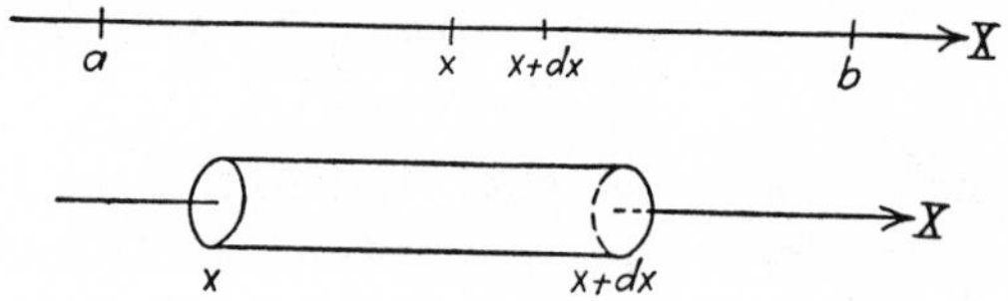

FIG. 25.

1) By *B*) the quantity of heat which enters at the left-hand end is

$$-k\frac{\partial\theta}{\partial x}\sigma\bigg]_x dt.$$

2) Similarly, the quantity of heat which flows out through the right-hand end is

$$-k\sigma\frac{\partial\theta}{\partial x}\bigg]_{x+dx} dt,$$

or, if we expand according to powers of dx and neglect higher powers of dx,

$$-k\sigma\frac{\partial\theta}{\partial x}\bigg]_x dt - \frac{\partial}{\partial x}\left(k\sigma\frac{\partial\theta}{\partial x}\right)\bigg]_x dtdx.$$

3) The quantity of heat which flows out through the cylindrical surface is, by C),

$$h(\theta - \Theta)l dx dt,$$

where l is the periphery of the cross-section. The total amount of heat which enters the element of the bar in the time dt is, therefore,

$$\left\{\frac{\partial}{\partial x}\left(k\,\sigma\,\frac{\partial \theta}{\partial x}\right) - h(\theta - \Theta)l\right\} dx dt.$$

This amount of heat increases the temperature of the element in the time dt by $\frac{\partial \theta}{\partial t}\, dt$ and, therefore, by A), is equal to $C\rho\sigma dx \frac{\partial \theta}{\partial t}\, dt$, where ρ is the density. Hence we obtain the partial differential equation

$$C\rho\sigma \frac{\partial \theta}{\partial t} = \frac{\partial}{\partial x}\left(k\,\sigma\,\frac{\partial \theta}{\partial x}\right) - h(\theta - \Theta)l. \tag{80}$$

If, moreover, heat is produced in the interior of the bar by electric currents or other sources of energy, let

$$A(x) dx dt$$

be the quantity of heat produced in the time dt in the interior of the element of the bar between the cross-sections, at x and $x + dx$. Then the partial differential equation (80) becomes

$$c\rho\sigma \frac{\partial \theta}{\partial x} = \frac{\partial}{\partial x}\left(k\sigma\,\frac{\partial \theta}{\partial x}\right) - h(\theta - \Theta)l + A(x).$$

In addition, the temperature θ has to satisfy certain boundary conditions, which are obtained as follows:

The amount of heat which leaves at a is, by C),

$$h(\theta - \Theta)\sigma\Big]_a dt. \tag{81}$$

On the other hand, the quantity of heat which flows through the cross-section $a + h$ in the direction of the negative x-axis in the time dt is, by B),

$$k\sigma \left.\frac{\partial \theta}{\partial x}\right]_{a+h} dt.$$

The limit of this expression as h approaches zero must be equal to the quantity (81). Hence we obtain the first boundary condition

$$h(\theta - \Theta) - k \left.\frac{\partial \theta}{\partial x}\right]_a = 0.$$

The same reasoning applied to the cross-section b gives us the second boundary condition

$$h(\theta - \Theta) + k \left.\frac{\partial \theta}{\partial x}\right]_b = 0.$$

c) The Special Case $\Theta \equiv 0$, $A \equiv 0$.—We now consider the differential equation under the assumption that

$$\Theta \equiv 0, \; A \equiv 0.$$

The differential equation is, then,

$$c\sigma\rho \frac{\partial \theta}{\partial t} = \frac{\partial}{\partial x}\left(k\sigma \frac{\partial \theta}{\partial x}\right) - hl\theta.$$

Put $\qquad c\sigma\rho = g(x),\; k\sigma = p(x),\; hl = -q(x),$

then $\qquad g(x) > 0,\; p(x) > 0,\; q(x) \leqq 0,$

and $\qquad q(x) = 0$ only when $h = 0$.

The differential equation of the problem now becomes

$$g \frac{\partial \theta}{\partial t} = \frac{\partial}{\partial x}\left(p \frac{\partial \theta}{\partial x}\right) + q\theta. \tag{82}$$

Put $$\frac{h(a)}{k(a)} = H_0 \geqq 0,\; \frac{h(b)}{k(b)} = H_1 \geqq 0,$$

then the boundary conditions become

$$\left\{\begin{aligned} \theta_x(a, t) - H_0\theta(a, t) &= 0 \\ \theta_x(b, t) + H_1\theta(b, t) &= 0 \end{aligned}\right. (t). \tag{83}$$

The initial condition is

$$\theta(x, 0) = f(x). \tag{84}$$

$f(x)$ is not entirely arbitrary, for from (83) we obtain for $t = 0$

$$\theta_x(a, 0) - H_0\theta(a, 0) = 0$$
$$\theta_x(b, 0) + H_1\theta(b, 0) = 0$$

whence

$$f'(a) - H_0f(a) = 0 \tag{85}$$
$$f'(b) + H_1f(b) = 0.$$

To solve (82), put

$$\theta = u(x)\phi(t).$$

Then, in the usual manner, (82) breaks up into the two ordinary differential equations:

$$\frac{d}{dx}\left(p\frac{du}{dx}\right) + qu + \lambda gu = 0 \tag{86}$$

$$\frac{d\phi}{dt} + \lambda\phi = 0, \tag{87}$$

while from (83) we get

$$u'(a) - H_0u(a) = 0 \tag{88}$$
$$u'(b) + H_1u(b) = 0$$

as boundary conditions on (86).

We have, as in our hypothesis (H_1), §72, that p is of class C', $p > 0$. In addition, we have $g > 0$ and $q \leqq 0$.

We shall now show that all of the characteristic constants of (86) are $\geqq 0$. This proof is analagous to that given in the proof of Theorem VIII.

We suppose that λ_0 is a characteristic constant and φ_0 a corresponding fundamental function. Therefore

$$L(\varphi_0) + \lambda_0\varphi_0 g = 0 \tag{89}$$

$$\varphi_0'(a) - H_0\varphi_0(a) = 0, \ \varphi_0'(b) + H_1\varphi_0(b) = 0 \tag{90}$$

and $\varphi_o \not\equiv 0$ is of class C''.

From (89) we obtain

$$\lambda_0 \int_a^b g\varphi_0^2 dx = -\int_a^b \varphi_0 L(\varphi_0) dx.$$

Integration by parts gives

$$\lambda_0 \int_a^b g\varphi_0^2 dx = -p(b)\varphi_0(b)\varphi_0'(b) + p(a)\varphi_0(a)\varphi_0'(a) + \int_a^b (p\varphi_0'^2 - q\varphi_0^2) dx,$$

which, by (90), reduces to

$$(91)\quad \lambda_0 \int_a^b g\varphi_0^2 dx = p(b)\varphi_0^2(b) H_1 + p(a)\varphi_0^2(a) H_0 + \int_a^b (p\varphi_0'^2 - q\varphi_0^2) dx.$$

Now $p(b)$, $\varphi_0^2(b)$, $p(a)$, $\varphi_0^2(a)$, and p are positive and not zero, and H_1, H_0, $\varphi_0'^2$, φ_0^2, $-q$ are $\geqq 0$. Whence, from (91), we conclude

$$\lambda_0 \geqq 0.$$

The equality will hold only when $H_1 = 0$, $H_0 = 0$, $q \equiv 0$ simultaneously. But $-q = hl$ and $l > 0$. Therefore $h \equiv 0$, which means that no heat escapes through the cylindrical surface. It also follows that no heat escapes through the ends for $h(a) = 0$, $h(b) = 0$, and hence the equality holds only when no heat escapes to the surrounding medium.

We can now construct the Green's function $K(x, t)$ and establish the equivalence of the boundary problem with the integral equation

$$u(x) = \lambda \int_a^b K(x, t) g(t) u(t) dt.$$

The kernel is not symmetric but the substitution

$$u(x)\sqrt{g(x)} = \bar{u}(x), \qquad K(x, t)\sqrt{g(x)g(t)} = \bar{K}(x, t)$$

transforms the problem into one with a symmetric kernel. This transformation leads the given differential equation into one in which $g \equiv 1$. We can then apply the results of the general Hilbert-Schmidt theory. Further, we can apply the Theorems V to VIII. We know then the existence of an infinitude of real characteristic constants each of index 1:

$$0 \leqq \lambda_1 < \lambda_2 < \lambda_3 < \cdot \cdot \cdot$$

with corresponding normalized fundamental functions

$$\psi_1(x), \psi_2(x), \psi_3(x), \cdot \cdot \cdot$$

We now turn to the solution of (87) for $\lambda = \lambda_n$. The solution is

$$\phi(t) = C_n e^{-\lambda_n t}$$

whence

$$\theta = C_n e^{-\lambda_n t} \psi_n(x)$$

is a solution of (82) which satisfies the boundary conditions but in general will not satisfy the initial conditions (84). Construct

$$\theta = \sum_{n=1}^{\infty} C_n e^{-\lambda_n t} \psi_n(x). \tag{92}$$

If we assume that (92) is convergent and admits one term-by-term partial differentiation with respect to t and two term-by-term partial differentiations with respect to x, then (92) satisfies (82) and (83), and (84) will be satisfied if

$$\sum_{n=1}^{\infty} C_n \psi_n(x) = f(x),$$

that is, if $f(x)$ can be expanded into a series of fundamental functions. That this is possible follows from Theorem VI when we take account of (85), provided f is of class C'', and C_n will be given by

$$C_n = (f\psi_n).$$

85. Application to an Example.—We consider now a special case where C, σ, ρ, k, h, l are constants and $a = 0$,

$b = 1$. Hence, p, q, g in (86) are constant. Put $\frac{q}{p} = -b^2$, and $\frac{\lambda g}{p}$ = constant, which we use as a new parameter and again designate by λ. Then (86) becomes

$$\frac{d^2u}{dx^2} + (\lambda - b^2)u = 0.$$

If we now assume that no heat escapes through the two end surfaces, then $H_0 = H_1 = 0$ and the boundary conditions become

$$u'(0) = 0,\ u'(1) = 0.$$

For this boundary problem the characteristic constants are easily found to be

$$\lambda_0 = b^2,\ \lambda_n = n^2\pi^2 + b^2,\ n = 1, 2, 3, \ldots$$

and the normalized fundamental functions are $\psi_0(x) = 1$, $\psi_n(x) = \sqrt{2} \cos n\pi x$.

For the further discussion we have to distinguish two cases:

Case I.—$b \neq 0$. Then all of the characteristic constants are positive. Therefore $\lambda = 0$ is not a characteristic constant and condition (H_2) is satisfied. Hence for this boundary problem we can construct a Green's function satisfying the following conditions:

A) K is continuous on $\left[01\right]$.

B) K is of class C'' on $\left[0t\right] \left[t1\right]$ separately.

$$\frac{d^2K}{dx^2} - b^2K = 0 \text{ on } \left[0t\right] \left[t1\right] \text{ separately.}$$

C) $K'(0) = 0$, $K'(1) = 0$.

D) $K'(t - 0) - K'(t + 0) = 1$.

The Green's function satisfying these conditions is easily found to be

$$K(x, t) = \begin{cases} \dfrac{\cosh bx \cosh b(1 - t)}{b \sinh b}, & 0 \leqq x \leqq t \\ \dfrac{\cosh bt \cosh b(1 - x)}{b \sinh b}, & t \leqq x \leqq 1. \end{cases}$$

It is symmetric and the boundary problem is equivalent to the integral equation

$$u(x) = \lambda \int_0^1 K(x, t)u(t)dt.$$

Theorem VI of this chapter assures us that every fundamental function of class C'' of this boundary problem, for which $u'(0) = 0$, $u'(1) = 0$ can be expanded into a cosine series

$$u(x) = \sum_{n=0}^{\infty} C_n \cos n\pi x$$

convergent on the interval $0 \leqq x \leqq 1$.

Case II.—$b = 0$. The boundary problem now becomes

$$\frac{d^2u}{dx^2} + \lambda u = 0,\ u'(0) = 0,\ u'(1) = 0, \tag{93}$$

for which the characteristic constants are

$$\lambda_0 = 0,\ \lambda_n = n^2\pi^2,\ n = 1, 2, 3, \ldots$$

with the corresponding normalized fundamental functions

$$\psi_0(x) = 1,\ \psi_n(x) = \sqrt{2} \cos n\pi x.$$

Let us consider

$$\bar{K}(x, t) = K(x, t) - \frac{1}{b^2}.$$

From the conditions on K we find that $\bar{K}$ satisfies the same conditions. Now let

$$\lim_{b \to 0} \bar{K}(x, t) = H(x, t).$$

We find

$$H(x, t) = \begin{cases} \dfrac{x^2 + t^2}{2} + \dfrac{1}{3} - t \equiv H_0, \; 0 \leqq x \leqq t \\ \dfrac{x^2 + t^2}{2} + \dfrac{1}{3} - x \equiv H_1, \; t \leqq x \leqq 1. \end{cases}$$

We see that $H(x, t)$ is symmetric and satisfies the following conditions:

A) H is continuous on $\left[01\right]$.

B) H is of class C'' on $\left[0t\right]\left[t1\right]$ separately.

$$\frac{d^2H}{dx^2} = 1 \text{ on } \left[0t\right]\left[t1\right] \text{ separately.}$$

C) $H'(0) = 0, H'(1) = 0.$

D) $H'(t - 0) - H'(t + 0) = 1.$

E) $\int_0^1 H(x, t)dt = 0, (t).$

We will now show that the boundary problem (93) is equivalent to

$$u(x) = \lambda \int_0^1 H(x, t)u(t)dt.$$

A) Multiply both members of $\dfrac{d^2u}{dx^2} = -\lambda u$ by $-H$, and both members of $\dfrac{d^2H}{dx^2} = 1$ by u and add. We obtain

$$\frac{d}{dx}(uH' - Hu') = \lambda uH + u,$$

whence

$$(uH' - Hu')\Big]_0^{t-0} + (uH' - Hu')\Big]_{t+0}^1 = \int_0^1 u(x)dx + \lambda \int_0^1 H(x, t)u(x)dx,$$

which, on account of (93) and condition C) on H, reduces to

$$(uH' - Hu')\Big]_{t+0}^{t-0} = \lambda \int_0^1 u(x)dx + \lambda \int_0^1 H(x, t)u(x)dx,$$

which, by conditions A) and D) on H, reduces to

$$u(t) = \lambda \int_0^1 u(x)dx + \lambda \int_0^1 H(x, t)u(x)dx.$$

Now from
$$\frac{d^2u}{dx^2} = -\lambda u(x)$$

we obtain, by integration,

$$-\int_0^1 u(x)dx = \int_0^1 \frac{d^2u}{dx^2}dx = \frac{du}{dx}\Big]_0^1 = 0.$$

Therefore

$$u(x) = \lambda \int_0^1 H(x, t)u(t)dt. \tag{94}$$

Thus we have shown that every solution of the boundary problem (93) with $\lambda \neq 0$ is a solution of the integral equation (94).

B) Conversely, suppose that u is continuous and

$$u(x) = \lambda \int_0^1 H(x, t)u(t)dt, \text{ and } \lambda \neq 0. \tag{95}$$

Then

$$u(x) = \lambda \int_0^x H_1(x, t)u(t)dt + \lambda \int_x^1 H_0(x, t)u(t)dt,$$

whence

$$u'(x) = \lambda \int_0^x H_1'(x, t)u(t)dt + \lambda \int_x^1 H_0'(x, t)u(t)dt \tag{96}$$

and

$$u''(x) = \lambda \int_0^x H_1''(x, t)u(t)dt + \lambda \int_x^1 H_0''(x, t)u(t)dt$$
$$+ \lambda H_1'(x, x)u(x) - \lambda H_0'(x, x)u(x).$$

But
$$H'' = 1 \text{ and } H'(t - 0) - H'(t + 0) = 1.$$

Therefore

$$u''(x) + \lambda u(x) = \lambda \int_0^1 u(t)dt.$$

But from (95)

$$\int_0^1 u(x)dx = \lambda \int_0^1 \int_0^1 H(x, t)u(t)dtdx$$
$$= \lambda \int_0^1 u(t)dt \int_0^1 H(x, t)dx = 0$$

on account of E). Therefore

$$u''(x) + \lambda u(x) = 0.$$

Further, from (96) we obtain

$$u'(0) = \lambda \int_0^1 H_0'(0, t)u(t)dt = 0$$
$$u'(1) = \lambda \int_0^1 H_1'(1, t)u(t)dt = 0,$$

on account of C). Therefore we have proved the equivalence of the boundary problem and integral equation.

86. General Theory of the Exceptional Case.—The method just applied to the special boundary problem (93) is a special case of the general method which Hilbert uses in the exceptional case where $\lambda = 0$ is a characteristic constant of the boundary problem:

(97) $$L(u) + \lambda u = 0$$

(98) $$R_0(u) = 0, \; R_1(u) = 0.$$

Let $\psi_0(x)$ be a normalized fundamental function belonging to $\lambda = 0$, so that ψ_0 is of class C'',

$$L(\psi_0) = 0, \; R_0(\psi_0) = 0, \; R_1(\psi_0) = 0,$$
$$\int_a^b \psi_0^2(x)dx = 1.$$

Then, according to Hilbert,[1] the equivalence of (97) (98) with an integral equation can be established as follows.

a) The Modified Green's Function.—

[1] Gött. Nach., 2, p. 219, 1904.

Theorem I.—*Under the above assumptions there exists one and only one function $H(x, t)$ satisfying as a function of x the following conditions:*

$A)$ H is continuous on $\left[ab\right]$.

$B)$ H is of class C'' on $\left[at\right]\left[tb\right]$ separately.

$$L(H) = \psi_0(x)\psi_0(t) \text{ on } \left[at\right]\left[tb\right] \text{ separately.}$$

$C)$ $R_0(H) = 0,\ R_1(H) = 0.$

$D)$ $H'(t - 0) - H'(t + 0) = \dfrac{1}{p\ (t)}.$

$E)$ $\displaystyle\int_a^b H(x, t)\psi_0(x)dx = 0,\ (t).$

$H(x, t)$ is called the modified Green's function. An outline of the proof of this theorem is as follows:

Let $u_0(x)$ be a particular solution of

$$L(u) = \psi_0(x)\psi_0(t),$$

then the general solution is

$$u = u_0(x) + \alpha\psi_0(x) + \beta V(x),$$

where V is a particular solution of $L(u) = 0$, independent of $\psi_0(x)$.

Now, by Green's theorem,

$$p(\psi_0 V' - \psi_0' V) = C$$

and we can select V in such a way that $C = 1$.

We now put

$$H(x, t) = \begin{cases} u_0(x) + \alpha_0\psi_0(x) + \beta_0 V(x),\ x \leqq t \\ u_0(x) + \alpha_1\psi_0(x) + \beta_1 V(x),\ t \leqq x. \end{cases}$$

Then

(99)
$$\begin{aligned} R_0(H) &= R_0(u_0) + \beta_0 R_0(V) = 0 \\ R_1(H) &= R_1(u_0) + \beta_1 R_1(V) = 0 \end{aligned}$$

determine β_0, β_1, since $R_0(V) \neq 0$, $R_1(V) \neq 0$.

The conditions A) to D) determine $\alpha_0 - \alpha_1$ and $\beta_0 - \beta_1$:

$$\alpha_0 - \alpha_1 = -V(t),\ \beta_0 - \beta_1 = \psi_0(t).$$

Since β_0, β_1 have already been determined, this furnishes an apparent contradiction, but it will be found that the determination of $\beta_0 - \beta_1$ is a consequence of the determination of β_0 and β_1 separately. Thus the conditions A) to D) determine β_0, β_1, α_0, α_1, except for an additive constant which is determined by E).

Theorem II. *The function $H(x, t)$ is symmetric.*—

For let

$$a \leqq t \leqq s \leqq b.$$

Multiply each member of

$$L\left\{ H(x, t) \right\} = \psi_0(x)\psi_0(t)$$

by $-H(x, s)$ and each member of

$$L\left\{ H(x, s) \right\} = \psi_0(x)\psi_0(s)$$

by $H(x, t)$. Add and integrate from a to t, t to s, s to b and add the results. We obtain

$$H(t, s) = H(s, t).$$

b) Hilbert's fundamental Theorem III now takes the form

Theorem III .—*If f is continuous, then the statements F is of class C'', $L(F) + f = 0$, $R_0(F) = 0$, $R_1(F) = 0$ imply and are implied by*

$$F(x) = \int_a^b H(x, t)f(t)dt.$$

Hence the expansion Theorem VI has to be modified as follows:

If F is of class C'', $R_0(F) = 0$, $R_1(F) = 0$,

then

$$F(x) = \sum C_\nu\psi_\nu(x)$$

where

$$C_\nu = (F\psi_\nu).$$

Applying this theorem when $F = u$, $f = \lambda u$, $\lambda \neq 0$ we obtain: if u is continuous, then the statements, u is of class C'', $L(u) + \lambda u = 0$, $R_0(u) = 0$, $R_1(u) = 0$, $\lambda \neq 0$, imply and are implied by

$$u(x) = \lambda \int_a^b H(x, t)u(t)dt.$$

This establishes the equivalence of the boundary problem and the integral equation.

87. Flow of Heat in a Ring.—The deductions which we made in the previous problem for a straight bar hold for any linear conductor if we take for the independent variable x the length of arc from a fixed point O. The results also hold if the ring is closed and the two end points A and B coincide. When A and B coincide, the boundary conditions are modified as shown later.

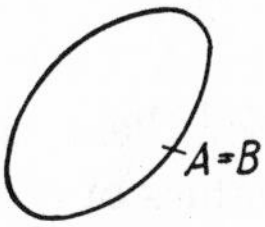

FIG. 26.

Let us suppose $\Theta \equiv 0$, $A \equiv 0$, total length of ring $= 1$. Then the boundary problem becomes

$$g\frac{\partial\theta}{\partial t} = \frac{\partial}{\partial x}\left(p\frac{\partial\theta}{\partial x}\right) + q\theta \tag{100}$$

$$\theta(0, t) = \theta(1, t); \; \theta_x(0, t) = \theta_x(1, t)$$

from the continuity of the temperature and of the flow of heat at A. Also

$$\theta(x, 0) = f(x), \text{ initial temperature.}$$

Put
$$\theta = u(x)\phi(t) \text{ in } (100).$$

We obtain

$$\frac{d^2u}{dx^2} + (\lambda - b^2)u = 0 \tag{101}$$

$$u(0) = u(1), \; u'(0) = u'(1). \tag{102}$$

Case I.—$\lambda - b^2 > 0$, say $\lambda - b^2 = \mu^2$.

The general solution of (101) is

$$u = \alpha \cos \mu x + \beta \sin \mu x.$$

The boundary conditions (102) give the following two equations for the determination of α, β.

$$\begin{aligned} &\alpha(\cos \mu - 1) + \beta \sin \mu = 0 \\ -&\alpha \sin \mu + \beta(\cos \mu - 1) = 0. \end{aligned} \tag{103}$$

These two equations are compatible for values of α, β not both zero, if and only if

$$\begin{vmatrix} \cos \mu - 1 & \sin \mu \\ -\sin \mu & \cos \mu - 1 \end{vmatrix} = 0.$$

That is, if
$$2(1 - \cos \mu) = 0,$$
whence
$$\mu = 2n\pi.$$

For these values of μ, equations (103) are satisfied by all values of α and β. For other values of μ, the only solution is the trivial one $u \equiv 0$. The characteristic constants are then

$$\lambda_n = b^2 + 4n^2\pi^2, \; n = 1, 2, \ldots,$$

each of index 2.

The corresponding fundamental functions are

$$u = \alpha \cos 2n\pi x + \beta \sin 2n\pi x.$$

Case II.—$\lambda - b^2 = 0$. Then $\lambda = b^2$ is the only characteristic constant. $u =$ constant is the only solution.

Case III.—$\lambda - b^2 < 0$. Then the only solution is the trivial one $u \equiv 0$. The characteristic constants of the problem are then $\lambda = b^2$, $\lambda_n = b^2 + 4n^2\pi^2$, $n = 1, 2, \ldots$ with normalized fundamental functions

$$1, \sqrt{2} \cos 2n\pi x, \sqrt{2} \sin 2n\pi x$$

and, since $\int_0^1 \cos 2n\pi x \sin 2n\pi x \, dx = 0$, the complete normalized orthogonal system of fundamental functions is

$$1, \sqrt{2} \cos 2n\pi x, \sqrt{2} \sin 2n\pi x$$

with the characteristic constants

$$b^2, b^2 + 4n^2\pi^2, b^2 + 4n^2\pi^2.$$

This seems to contradict Theorem VII of §75, which stated that the index is always 1. The assumptions of this theorem were, however, that the boundary conditions were of the form

$$Au(a) + Bu'(a) = 0$$
$$Cu(b) + Du'(b) = 0.$$

The boundary conditions of the above problem are of a more general type:

$$A_0u(a) + B_0u'(a) + C_0u(b) + D_0u'(b) = 0$$
$$A_1u(a) + B_1u'(a) + C_1u(b) + D_1u'(b) = 0,$$

which explains the apparent contradiction.

If $b \neq 0$, (H_2) is satisfied and we can proceed to construct Green's function:

$$K(x, t) = \begin{cases} \dfrac{\cosh b(t - x - \frac{1}{2})}{2b \sinh \frac{b}{2}}, & x \leq t \\[2ex] \dfrac{\cosh b(x - t - \frac{1}{2})}{2b \sinh \frac{b}{2}}, & x \geq t \end{cases}$$

and the boundary problem is equivalent to the integral equation

$$u(x) = \lambda \int_a^b K(x, t)u(t)dt.$$

The expansion theorem can now be stated as follows: If F is of class C'', $F(0) = F(1)$ and $F'(0) = F'(1)$ then

$$F(x) = A_0 + \sum (A_n \cos 2n\pi x + B_n \sin 2n\pi x).$$

This is an ordinary Fourier series for the expansion of $F(x)$ for the interval $\left[01\right]$.

If $b = 0$, then (H_2) is not satisfied. Then, as in §86, we construct[1] a modified kernel:

$$H(x, t) = \begin{cases} \frac{1}{2}(t - x - \frac{1}{2})^2 + \frac{1}{24}, & x \leq t \\ \frac{1}{2}(x - t - \frac{1}{2})^2 + \frac{1}{24}, & t \leq x. \end{cases}$$

[1] See KNESER, "Integralgleichungen," §7.

88. Stationary Flow of Heat Produced by an Interior Source.—We have so far been considering the problem of the flow of heat under the assumption that no heat was produced in the interior of the conductor: $A \equiv 0$.

a) Solution of the Problem of Stationary Flow of Heat.—Let us now consider the case $A \not\equiv 0$. The equations of the problem now are

$$g\frac{\partial\theta}{\partial t} = \frac{\partial}{\partial x}\left(p\frac{\partial\theta}{\partial x}\right) + q\theta + A(x)$$

$$H_0\theta - \frac{\partial\theta}{\partial x} = 0 \text{ for } x = a,\ H_0 \geqq 0$$

$$H_1\theta + \frac{\partial\theta}{\partial x} = 0 \text{ for } x = b,\ H_1 \geqq 0.$$

Instead of the initial condition $\theta(x, 0) = f(x)$, we require that the flow of heat shall be stationary, that is, independent of the time, that is, $\theta = w(x)$ and, therefore, $\frac{\partial\theta}{\partial t} \equiv 0$. Then the equations of the problem are

$$(104)\quad \frac{d}{dx}\left(p\frac{dw}{dx}\right) + qw + A(x) = 0 \quad \left[L(w) + A = 0\right]$$

$$(105)\quad \begin{aligned} R_0(w) &\equiv H_0w(a) - w'(a) = 0 \\ R_1(w) &\equiv H_1w(b) + w'(b) = 0. \end{aligned}$$

This is a non-homogeneous boundary problem. From Hilbert's third fundamental theorem we can write down at once the following:

If A is continuous, then the statements, w is of class C'', $L(w) + A = 0$, $R_0(w) = 0$, $R_1(w) = 0$, imply and are implied by

$$(106)\quad w(x) = \int_a^b K(x, t)A(t)dt.$$

That is, the boundary problem (104) (105) has one and only one solution given by (106). Thus we have the theorem: *Every continuous source of heat $A(x)$ produces one and only one stationary flow of heat expressed by* (106).

b) Physical Interpretation of Green's Function.—From (106) Kneser obtains a physical interpretation for Green's function.

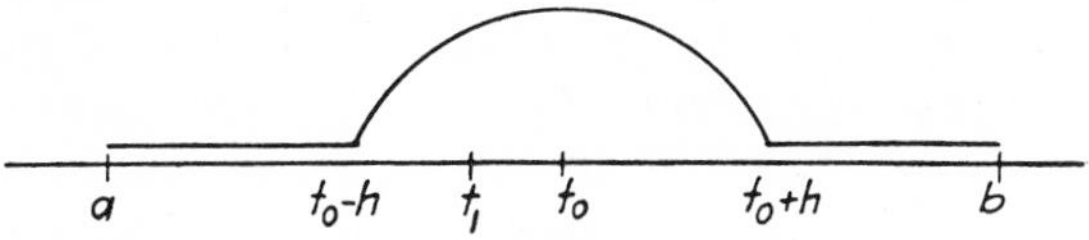

FIG. 27.

Take t_0 between a and b and let us suppose that

$$A(x)\begin{cases} \equiv 0 \text{ for } a \leqq x \leqq t_0 - h,\ t_0 + h \leqq x \leqq b \\ \geqq 0 \text{ for } t_0 - h \leqq x \leqq t_0 + h. \end{cases}$$

Then, from the first mean value theorem for definite integrals, there exists a t_1, $t_0 - h \leqq t_1 \leqq t_0 + h$ such that

$$w(x, h) = \int_{t_0-h}^{t_0+h} K(x, t)A(t)dt = K(x, t_1)\int_{t_0-h}^{t_0+h} A(t)dt.$$

Now $A(x)dx$ is the quantity of heat produced in an element dx in unit of time, and hence $\int_a^b A(t)dt$ is the total amount of heat produced in the bar in unit time. This is called the *strength of source of heat.*

Let us suppose that the strength of the source of heat is 1:

$$\int_{t_0-h}^{t_0+h} A(t)dt = 1. \tag{107}$$

Let now $h \to 0$ and let at the same time $A(x)$, which depends upon h, so vary that (107) remains satisfied. Then

$$K(x, t_0) = \lim_{h \to 0} w(x, h).$$

Thus we have the theorem: *In the general case, Green's function $K(x, t)$ represents the stationary temperature produced by a point source of strength unity placed at $x = t$.*

89. Direct Computation of the Characteristic Constants and Fundamental Functions.—In all of the examples in which we have been able actually to determine the characteristic constants and the fundamental functions, we have determined the latter by means of a boundary problem.

The question arises: How could we directly determine the characteristic constants, and the fundamental functions for an integral equation which does not correspond to a boundary problem? We might obtain the characteristic constants by the solution of the equation $D(\lambda) = 0$, and determine the index and the fundamental functions by means of the Fredholm minors. For some simple kernels this is easily done, but in the general case this method is hardly practicable.

Another direct method has been developed by Schmidt ("Diss.," pages 18 to 21) and Kneser ("Integralgleichungen," pages 190 to 197), at least for a symmetric kernel.

For simplicity, let us suppose that we knew *a priori* that all of the characteristic constants are positive and of index 1.

a) Determination of λ_1.—It was established in the general theory that if $0 < \lambda_1 < \lambda_2 < \ . \ . \ .$ were the characteristic constants for $K(x, t)$ and $\psi_1(x), \psi_2(x), \ . \ . \ .$ the corresponding complete normalized orthogonal system of fundamental functions, then the kernel $K_2(x, t)$ had the same fundamental functions and they belong to the characteristic constants $0 < \lambda_1^2 < \lambda_2^2 < \ . \ . \ .$, for which we write $\mu_1 < \mu_2 < \ . \ . \ . \ .$ Now we have the following expansion for the logarithmic derivative of Fredholm's determinant $D(\lambda)$ derived from the kernel $K(x, t)$:

$$\frac{D'(\lambda)}{D(\lambda)} = -\sum_{n=0}^{\infty} U_{n+1}\lambda^n \qquad [(14), \S 54].$$

where $U_n = \int_a^b K_n(x, x)dx$, convergent for sufficiently small values of λ.

Hence if we call $D_2(\mu)$ the Fredholm determinant for $K_2(x, t)$, we obtain the corresponding expression

$$\frac{D'_2(\mu)}{D_2(\mu)} = -\sum_{n=0}^{\infty} U_{2n+2}\mu^n. \tag{108}$$

Now $\frac{D'_2(\mu)}{D_2(\mu)}$ is a meromorphic function with simple poles, the poles being the roots of $D_2(\mu) = 0$. Arrange the roots $0 < \mu_1 < \mu_2 < \mu_3 < \ . \ . \ .$ in order of magnitude. Hence the circle of convergence of the expression (106) for $\frac{D'_2}{D_2}$ passes through μ_1 and hence the radius of convergence R is equal to μ_1.

Now there is a theorem[1] on power series to the effect that the radius of convergence R for the series $\sum A_n z^n$ is given by

$$\lim_{n \to \infty} \frac{A_n}{A_{n+1}} = R,$$

provided this ratio has a determinate limit as $n \to \infty$.

This theorem applied to the present problem gives

$$\lim_{n \to \infty} \frac{U_{2n}}{U_{2n+2}} = \mu_1,$$

provided the limit exists. That this ratio has a determinate limit follows from the monotony principle applied to the inequality

$$\frac{U_{2n-2}}{U_{2n}} \geqq \frac{U_{2n}}{U_{2n+2}} > 0$$

which we proved in §58.

b) *Computation of the Fundamental Function* $\psi_1(x)$.—From

$$K_{2n+2}(x, y) = \int_a^b K_2(x, t) K_{2n}(t, y) dt$$

we obtain

$$(109) \quad \mu_1^{n+1} K_{2n+2}(x, y) = \mu_1 \int_a^b K_2(x, t) \Big(\mu_1^n K_{2n}(t, y) \Big) dt.$$

Now Schmidt proves in his existence proof for the characteristic constants that

$$\lim_{n \to \infty} \mu_1^n K_{2n}(x, y) = f(x, y)$$

[1] HARKNESS and MORLEY, "Theory of Functions," §76.

a definite limit function which is approached uniformly with respect to x and y in the region R. Therefore, by passing to the limit in (109), we find

$$f(x, y) = \mu_1 \int_a^b K_2(x, t) f(t, y) dt.$$

Further Schmidt proves that $f(x, x) \not\equiv 0$. Hence, if for some quantity c we have $f(c, c) \neq 0$, then the function $f(x, c) = \varphi(x) \not\equiv 0$ and, therefore, $\varphi(x)$ is a fundamental function belonging to μ_1 since it satisfies the equation.

This function $\varphi(x)$, normalized in the usual way, gives us the fundamental function $\psi_1(x)$ belonging to λ_1.

c) Computation of the Other Characteristic Constants and Fundamental Functions.—The other characteristic constants and fundamental functions can be obtained successively as follows. Let

$$\text{(110)} \qquad K_2'(x, t) = K_2(x, t) - \frac{\psi_1(x)\psi_1(t)}{\mu_1}$$

be a new kernel, then we can show that its characteristic constants and fundamental functions are

$$\lambda_2^2 < \lambda_3^2 < \ . \ . \ . \ ; \psi_2(x), \psi_3(x), \ . \ . \ . \ .$$

For from (110) we obtain

$$\text{(111)} \quad \int_a^b K_2'(x, t)\psi_\nu(t) dt = \int_a^b K_2(x, t)\psi_\nu(t) dt + \frac{\psi_1(x)}{\mu_1} \delta_{1\nu},$$

where $\delta_{1\nu}$ is the Kronecker symbol. When $\nu = 1$, (111) gives

$$\int_a^b K_2'(x, t)\psi_1(t) dt = 0.$$

When $\nu \neq 1$, (111) gives

$$\psi_\nu(x) = \mu_\nu \int_a^b K_2'(x, t)\psi_\nu(t) dt,$$

which proves the above statement.

Applying now to the kernel $K_2'(x, t)$ the method described under *a*) and *b*), we obtain λ_2^2 and $\psi_2(x)$, and so on.

INDEX

CATALOGUE OF DOVER BOOKS

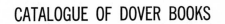

MATHEMATICS—INTERMEDIATE TO ADVANCED

General

INTRODUCTION TO APPLIED MATHEMATICS, Francis D. Murnaghan. A practical and thoroughly sound introduction to a number of advanced branches of higher mathematics. Among the selected topics covered in detail are: vector and matrix analysis, partial and differential equations, integral equations, calculus of variations, Laplace transform theory, the vector triple product, linear vector functions, quadratic and bilinear forms, Fourier series, spherical harmonics, Bessel functions, the Heaviside expansion formula, and many others. Extremely useful book for graduate students in physics, engineering, chemistry, and mathematics. Index. 111 study exercises with answers. 41 illustrations. ix + 389pp. 5⅜ x 8½.
S1042 Paperbound **$2.00**

OPERATIONAL METHODS IN APPLIED MATHEMATICS, H. S. Carslaw and J. C. Jaeger. Explanation of the application of the Laplace Transformation to differential equations, a simple and effective substitute for more difficult and obscure operational methods. Of great practical value to engineers and to all workers in applied mathematics. Chapters on: Ordinary Linear Differential Equations with Constant Coefficients;; Electric Circuit Theory; Dynamical Applications; The Inversion Theorem for the Laplace Transformation; Conduction of Heat; Vibrations of Continuous Mechanical Systems; Hydrodynamics; Impulsive Functions; Chains of Differential Equations; and other related matters. 3 appendices. 153 problems, many with answers. 22 figures. xvi + 359pp. 5⅜ x 8½.
S1011 Paperbound **$2.25**

APPLIED MATHEMATICS FOR RADIO AND COMMUNICATIONS ENGINEERS, C. E. Smith. No extraneous material here!—only the theories, equations, and operations essential and immediately useful for radio work. Can be used as refresher, as handbook of applications and tables, or as full home-study course. Ranges from simplest arithmetic through calculus, series and wave forms, hyperbolic trigonometry, simultaneous equations in mesh circuits, etc. Supplies applications right along with each math topic discussed. 22 useful tables of functions, formulas, logs, etc. Index. 166 exercises, 140 examples, all with answers. 95 diagrams. Bibliography. x + 336pp. 5⅜ x 8.
S141 Paperbound **$1.75**

Algebra, group theory, determinants, sets, matrix theory

ALGEBRAS AND THEIR ARITHMETICS, L. E. Dickson. Provides the foundation and background necessary to any advanced undergraduate or graduate student studying abstract algebra. Begins with elementary introduction to linear transformations, matrices, field of complex numbers; proceeds to order, basal units, modulus, quaternions, etc.; develops calculus of linears sets, describes various examples of algebras including invariant, difference, nilpotent, semi-simple. "Makes the reader marvel at his genius for clear and profound analysis," Amer. Mathematical Monthly. Index. xii + 241pp. 5⅜ x 8.
S616 Paperbound **$1.50**

THE THEORY OF EQUATIONS WITH AN INTRODUCTION TO THE THEORY OF BINARY ALGEBRAIC FORMS, W. S. Burnside and A. W. Panton. Extremely thorough and concrete discussion of the theory of equations, with extensive detailed treatment of many topics curtailed in later texts. Covers theory of algebraic equations, properties of polynomials, symmetric functions, derived functions, Horner's process, complex numbers and the complex variable, determinants and methods of elimination, invariant theory (nearly 100 pages), transformations, introduction to Galois theory, Abelian equations, and much more. Invaluable supplementary work for modern students and teachers. 759 examples and exercises. Index in each volume. Two volume set. Total of xxiv + 604pp. 5⅜ x 8.
S714 Vol I Paperbound **$1.85**
S715 Vol II Paperbound **$1.85**
The set **$3.70**

COMPUTATIONAL METHODS OF LINEAR ALGEBRA, V. N. Faddeeva, translated by **C. D. Benster.** First English translation of a unique and valuable work, the only work in English presenting a systematic exposition of the most important methods of linear algebra—classical and contemporary. Shows in detail how to derive numerical solutions of problems in mathematical physics which are frequently connected with those of linear algebra. Theory as well as individual practice. Part I surveys the mathematical background that is indispensable to what follows. Parts II and III, the conclusion, set forth the most important methods of solution, for both exact and iterative groups. One of the most outstanding and valuable features of this work is the 23 tables, double and triple checked for accuracy. These tables will not be found elsewhere. Author's preface. Translator's note. New bibliography and index. x + 252pp. 5⅜ x 8.
S424 Paperbound **$1.95**

ALGEBRAIC EQUATIONS, E. Dehn. Careful and complete presentation of Galois' theory of algebraic equations; theories of Lagrange and Galois developed in logical rather than historical form, with a more thorough exposition than in most modern books. Many concrete applications and fully-worked-out examples. Discusses basic theory (very clear exposition of the symmetric group); isomorphic, transitive, and Abelian groups; applications of Lagrange's and Galois' theories; and much more. Newly revised by the author. Index. List of Theorems. xi + 208pp. 5⅜ x 8.
S697 Paperbound **$1.45**

THEORY AND APPLICATIONS OF FINITE GROUPS, G. A. Miller, H. F. Blichfeldt, L. E. Dickson. Unusually accurate and authoritative work, each section prepared by a leading specialist: Miller on substitution and abstract groups, Blichfeldt on finite groups of linear homogeneous transformations, Dickson on applications of finite groups. Unlike more modern works, this gives the concrete basis from which abstract group theory arose. Includes Abelian groups, prime-power groups, isomorphisms, matrix forms of linear transformations, Sylow groups, Galois' theory of algebraic equations, duplication of a cube, trisection of an angle, etc. 2 Indexes. 267 problems. xvii + 390pp. 5⅜ x 8. S216 Paperbound **$2.00**

THE THEORY OF DETERMINANTS, MATRICES, AND INVARIANTS, H. W. Turnbull. Important study includes all salient features and major theories. 7 chapters on determinants and matrices cover fundamental properties, Laplace identities, multiplication, linear equations, rank and differentiation, etc. Sections on invariants gives general properties, symbolic and direct methods of reduction, binary and polar forms, general linear transformation, first fundamental theorem, multilinear forms. Following chapters study development and proof of Hilbert's Basis Theorem, Gordan-Hilbert Finiteness Theorem, Clebsch's Theorem, and include discussions of apolarity, canonical forms, geometrical interpretations of algebraic forms, complete system of the general quadric, etc. New preface and appendix. Bibliography. xviii + 374pp. 5⅜ x 8. S699 Paperbound **$2.25**

AN INTRODUCTION TO THE THEORY OF CANONICAL MATRICES, H. W. Turnbull and A. C. Aitken. All principal aspects of the theory of canonical matrices, from definitions and fundamental properties of matrices to the practical applications of their reduction to canonical form. Beginning with matrix multiplications, reciprocals, and partitioned matrices, the authors go on to elementary transformations and bilinear and quadratic forms. Also covers such topics as a rational canonical form for the collineatory group, congruent and conjunctive transformation for quadratic and hermitian forms, unitary and orthogonal transformations, canonical reduction of pencils of matrices, etc. Index. Appendix. Historical notes at chapter ends. Bibliographies. 275 problems. xiv + 200pp. 5⅜ x 8. S177 Paperbound **$1.55**

A TREATISE ON THE THEORY OF DETERMINANTS, T. Muir. Unequalled as an exhaustive compilation of nearly all the known facts about determinants up to the early 1930's. Covers notation and general properties, row and column transformation, symmetry, compound determinants, adjugates, rectangular arrays and matrices, linear dependence, gradients, Jacobians, Hessians, Wronskians, and much more. Invaluable for libraries of industrial and research organizations as well as for student, teacher, and mathematician; very useful in the field of computing machines. Revised and enlarged by W. H. Metzler. Index. 485 problems and scores of numerical examples. iv + 766pp. 5⅜ x 8. S670 Paperbound **$3.00**

THEORY OF DETERMINANTS IN THE HISTORICAL ORDER OF DEVELOPMENT, Sir Thomas Muir. Unabridged reprinting of this complete study of 1,859 papers on determinant theory written between 1693 and 1900. Most important and original sections reproduced, valuable commentary on each. No other work is necessary for determinant research: all types are covered—each subdivision of the theory treated separately; all papers dealing with each type are covered; you are told exactly what each paper is about and how important its contribution is. Each result, theory, extension, or modification is assigned its own identifying numeral so that the full history may be more easily followed. Includes papers on determinants in general, determinants and linear equations, symmetric determinants, alternants, recurrents, determinants having invariant factors, and all other major types. "A model of what such histories ought to be," NATURE. "Mathematicians must ever be grateful to Sir Thomas for his monumental work," AMERICAN MATH MONTHLY. Four volumes bound as two. Indices. Bibliographies. Total of lxxxiv + 1977pp. 5⅜ x 8. S672-3 The set, Clothbound **$12.50**

Calculus and function theory, Fourier theory, infinite series, calculus of variations, real and complex functions

FIVE VOLUME "THEORY OF FUNCTIONS' SET BY KONRAD KNOPP

This five-volume set, prepared by Konrad Knopp, provides a complete and readily followed account of theory of functions. Proofs are given concisely, yet without sacrifice of completeness or rigor. These volumes are used as texts by such universities as M.I.T., University of Chicago, N. Y. City College, and many others. "Excellent introduction . . . remarkably readable, concise, clear, rigorous," JOURNAL OF THE AMERICAN STATISTICAL ASSOCIATION.

ELEMENTS OF THE THEORY OF FUNCTIONS, Konrad Knopp. This book provides the student with background for further volumes in this set, or texts on a similar level. Partial contents: foundations, system of complex numbers and the Gaussian plane of numbers, Riemann sphere of numbers, mapping by linear functions, normal forms, the logarithm, the cyclometric functions and binomial series. "Not only for the young student, but also for the student who knows all about what is in it," MATHEMATICAL JOURNAL. Bibliography. Index. 140pp. 5⅜ x 8. S154 Paperbound **$1.35**

THEORY OF FUNCTIONS, PART I, Konrad Knopp. With volume II, this book provides coverage of basic concepts and theorems. Partial contents: numbers and points, functions of a complex variable, integral of a continuous function, Cauchy's integral theorem, Cauchy's integral formulae, series with variable terms, expansion of analytic functions in power series, analytic continuation and complete definition of analytic functions, entire transcendental functions, Laurent expansion, types of singularities. Bibliography. Index. vii + 146pp. 5⅜ x 8. S156 Paperbound **$1.35**

THEORY OF FUNCTIONS, PART II, Konrad Knopp. Application and further development of general theory, special topics. Single valued functions, entire, Weierstrass, Meromorphic functions. Riemann surfaces. Algebraic functions. Analytical configuration, Riemann surface. Bibliography. Index. x + 150pp. 5⅜ x 8. S157 Paperbound **$1.35**

PROBLEM BOOK IN THE THEORY OF FUNCTIONS, VOLUME 1, Konrad Knopp. Problems in elementary theory, for use with Knopp's THEORY OF FUNCTIONS, or any other text, arranged according to increasing difficulty. Fundamental concepts, sequences of numbers and infinite series, complex variable, integral theorems, development in series, conformal mapping. 182 problems. Answers. viii + 126pp. 5⅜ x 8. S158 Paperbound **$1.35**

PROBLEM BOOK IN THE THEORY OF FUNCTIONS, VOLUME 2, Konrad Knopp. Advanced theory of functions, to be used either with Knopp's THEORY OF FUNCTIONS, or any other comparable text. Singularities, entire & meromorphic functions, periodic, analytic, continuation, multiple-valued functions, Riemann surfaces, conformal mapping. Includes a section of additional elementary problems. "The difficult task of selecting from the immense material of the modern theory of functions the problems just within the reach of the beginner is here masterfully accomplished," AM. MATH. SOC. Answers. 138pp. 5⅜ x 8. S159 Paperbound **$1.35**

A COURSE IN MATHEMATICAL ANALYSIS, Edouard Goursat. Trans. by E. R. Hedrick, O. Dunkel. Classic study of fundamental material thoroughly treated. Exceptionally lucid exposition of wide range of subject matter for student with 1 year of calculus. Vol. 1: Derivatives and Differentials, Definite Integrals, Expansion in Series, Applications to Geometry. Problems. Index. 52 illus. 556pp. Vol. 2, Part I: Functions of a Complex Variable, Conformal Representations, Doubly Periodic Functions, Natural Boundaries, etc. Problems. Index. 38 illus. 269pp. Vol. 2, Part 2: Differential Equations, Cauchy-Lipschitz Method, Non-linear Differential Equations, Simultaneous Equations, etc. Problems. Index. 308pp. 5⅜ x 8.
Vol. 1 S554 Paperbound **$2.50**
Vol. 2 part 1 S555 Paperbound **$1.85**
Vol. 2 part 2 S556 Paperbound **$1.85**
3 vol. set **$6.20**

MODERN THEORIES OF INTEGRATION, H. Kestelman. Connected and concrete coverage, with fully-worked-out proofs for every step. Ranges from elementary definitions through theory of aggregates, sets of points, Riemann and Lebesgue integration, and much more. This new revised and enlarged edition contains a new chapter on Riemann-Stieltjes integration, as well as a supplementary section of 186 exercises. Ideal for the mathematician, student, teacher, or self-studier. Index of Definitions and Symbols. General Index. Bibliography. x + 310pp. 5⅝ x 8⅜. S572 Paperbound **$2.25**

THEORY OF MAXIMA AND MINIMA, H. Hancock. Fullest treatment ever written; only work in English with extended discussion of maxima and minima for functions of 1, 2, or n variables, problems with subsidiary constraints, and relevant quadratic forms. Detailed proof of each important theorem. Covers the Scheeffer and von Dantscher theories, homogeneous quadratic forms, reversion of series, fallacious establishment of maxima and minima, etc. Unsurpassed treatise for advanced students of calculus, mathematicians, economists, statisticians. Index. 24 diagrams. 39 problems, many examples. 193pp. 5⅜ x 8. S665 Paperbound **$1.50**

AN ELEMENTARY TREATISE ON ELLIPTIC FUNCTIONS, A. Cayley. Still the fullest and clearest text on the theories of Jacobi and Legendre for the advanced student (and an excellent supplement for the beginner). A masterpiece of exposition by the great 19th century British mathematician (creator of the theory of matrices and abstract geometry), it covers the addition-theory, Landen's theorem, the 3 kinds of elliptic integrals, transformations, the q-functions, reduction of a differential expression, and much more. Index. xii + 386pp. 5⅜ x 8. S728 Paperbound **$2.00**

THE APPLICATIONS OF ELLIPTIC FUNCTIONS, A. G. Greenhill. Modern books forego detail for sake of brevity—this book offers complete exposition necessary for proper understanding, use of elliptic integrals. Formulas developed from definite physical, geometric problems; examples representative enough to offer basic information in widely useable form. Elliptic integrals, addition theorem, algebraical form of addition theorem, elliptic integrals of 2nd, 3rd kind, double periodicity, resolution into factors, series, transformation, etc. Introduction. Index. 25 illus. xi + 357pp. 5⅜ x 8. S603 Paperbound **$1.75**

THE THEORY OF FUNCTIONS OF REAL VARIABLES, James Pierpont. A 2-volume authoritative exposition, by one of the foremost mathematicians of his time. Each theorem stated with all conditions, then followed by proof. No need to go through complicated reasoning to discover conditions added without specific mention. Includes a particularly complete, rigorous presentation of theory of measure; and Pierpont's own work on a theory of Lebesgue integrals, and treatment of area of a curved surface. Partial contents, Vol. 1: rational numbers, exponentials, logarithms, point aggregates, maxima, minima, proper integrals, improper integrals, multiple proper integrals, continuity, discontinuity, indeterminate forms. Vol. 2: point sets, proper integrals, series, power series, aggregates, ordinal numbers, discontinuous functions, sub-, infra-uniform convergence, much more. Index. 95 illustrations. 1229pp. 5⅜ x 8. S558-9, 2 volume set, paperbound **$5.20**

FUNCTIONS OF A COMPLEX VARIABLE, James Pierpont. Long one of best in the field. A thorough treatment of fundamental elements, concepts, theorems. A complete study, rigorous, detailed, with carefully selected problems worked out to illustrate each topic. Partial contents: arithmetical operations, real term series, positive term series, exponential functions, integration, analytic functions, asymptotic expansions, functions of Weierstrass, Legendre, etc. Index. List of symbols. 122 illus. 597pp. 5⅜ x 8. S560 Paperbound **$2.45**

MODERN OPERATIONAL CALCULUS: WITH APPLICATIONS IN TECHNICAL MATHEMATICS, N. W. McLachlan. An introduction to modern operational calculus based upon the Laplace transform, applying it to the solution of ordinary and partial differential equations. For physicists, engineers, and applied mathematicians. Partial contents: Laplace transform, theorems or rules of the operational calculus, solution of ordinary and partial linear differential equations with constant coefficients, evaluation of integrals and establishment of mathematical relationships, derivation of Laplace transforms of various functions, etc. Six appendices deal with Heaviside's unit function, etc. Revised edition. Index. Bibliography. xiv + 218pp. 5⅜ x 8½. S192 Paperbound **$1.75**

ADVANCED CALCULUS, E. B. Wilson. An unabridged reprinting of the work which continues to be recognized as one of the most comprehensive and useful texts in the field. It contains an immense amount of well-presented, fundamental material, including chapters on vector functions, ordinary differential equations, special functions, calculus of variations, etc., which are excellent introductions to these areas. For students with only one year of calculus, more than 1300 exercises cover both pure math and applications to engineering and physical problems. For engineers, physicists, etc., this work, with its 54 page introductory review, is the ideal reference and refresher. Index. ix + 566pp. 5⅜ x 8. S504 Paperbound **$2.45**

ASYMPTOTIC EXPANSIONS, A. Erdélyi. The only modern work available in English, this is an unabridged reproduction of a monograph prepared for the Office of Naval Research. It discusses various procedures for asymptotic evaluation of integrals containing a large parameter and solutions of ordinary linear differential equations. Bibliography of 71 items. vi + 108pp. 5⅜ x 8. S318 Paperbound **$1.35**

INTRODUCTION TO ELLIPTIC FUNCTIONS: with applications, F. Bowman. Concise, practical introduction to elliptic integrals and functions. Beginning with the familiar trigonometric functions, it requires nothing more from the reader than a knowledge of basic principles of differentiation and integration. Discussion confined to the Jacobian functions. Enlarged bibliography. Index. 173 problems and examples. 56 figures, 4 tables. 115pp. 5⅜ x 8. S922 Paperbound **$1.25**

ON RIEMANN'S THEORY OF ALGEBRAIC FUNCTIONS AND THEIR INTEGRALS: A SUPPLEMENT TO THE USUAL TREATISES, Felix Klein. Klein demonstrates how the mathematical ideas in Riemann's work on Abelian integrals can be arrived at by thinking in terms of the flow of electric current on surfaces. Intuitive explanations, not detailed proofs given in an extremely clear exposition, concentrating on the kinds of functions which can be defined on Riemann surfaces. Also useful as an introduction to the origins of topological problems. Complete and unabridged. Approved translation by Frances Hardcastle. New introduction. 43 figures. Glossary. xii + 76pp. 5⅜ x 8½. S1072 Paperbound **$1.25**

COLLECTED WORKS OF BERNHARD RIEMANN. This important source book is the first to contain the complete text of both 1892 Werke and the 1902 supplement, unabridged. It contains 31 monographs, 3 complete lecture courses, 15 miscellaneous papers, which have been of enormous importance in relativity, topology, theory of complex variables, and other areas of mathematics. Edited by R. Dedekind, H. Weber, M. Noether, W. Wirtinger. German text. English introduction by Hans Lewy. 690pp. 5⅜ x 8. S226 Paperbound **$3.75**

THE TAYLOR SERIES, AN INTRODUCTION TO THE THEORY OF FUNCTIONS OF A COMPLEX VARIABLE, P. Dienes. This book investigates the entire realm of analytic functions. Only ordinary calculus is needed, except in the last two chapters. Starting with an introduction to real variables and complex algebra, the properties of infinite series, elementary functions, complex differentiation and integration are carefully derived. Also biuniform mapping, a thorough two part discussion of representation and singularities of analytic functions, overconvergence and gap theorems, divergent series, Taylor series on its circle of convergence, divergence and singularities, etc. Unabridged, corrected reissue of first edition. Preface and index. 186 examples, many fully worked out. 67 figures. xii + 555pp. 5⅜ x 8. S391 Paperbound **$2.75**

INTRODUCTION TO BESSEL FUNCTIONS, Frank Bowman. A rigorous self-contained exposition providing all necessary material during the development, which requires only some knowledge of calculus and acquaintance with differential equations. A balanced presentation including applications and practical use. Discusses Bessel Functions of Zero Order, of Any Real Order; Modified Bessel Functions of Zero Order; Definite Integrals; Asymptotic Expansions; Bessel's Solution to Kepler's Problem; Circular Membranes; much more. "Clear and straightforward . . . useful not only to students of physics and engineering, but to mathematical students in general," Nature. 226 problems. Short tables of Bessel functions. 27 figures. Index. x + 135pp. 5⅜ x 8. S462 Paperbound **$1.35**

Differential equations, ordinary and partial; integral equations

INTRODUCTION TO THE DIFFERENTIAL EQUATIONS OF PHYSICS, L. Hopf. Especially valuable to the engineer with no math beyond elementary calculus. Emphasizing intuitive rather than formal aspects of concepts, the author covers an extensive territory. Partial contents: Law of causality, energy theorem, damped oscillations, coupling by friction, cylindrical and spherical coordinates, heat source, etc. Index. 48 figures. 160pp. 5⅜ x 8.
S120 Paperbound **$1.25**

INTRODUCTION TO THE THEORY OF LINEAR DIFFERENTIAL EQUATIONS, E. G. Poole. Authoritative discussions of important topics, with methods of solution more detailed than usual, for students with background of elementary course in differential equations. Studies existence theorems, linearly independent solutions; equations with constant coefficients; with uniform analytic coefficients; regular singularities; the hypergeometric equation; conformal representation; etc. Exercises. Index. 210pp. 5⅜ x 8. S629 Paperbound **$1.65**

DIFFERENTIAL EQUATIONS FOR ENGINEERS, P. Franklin. Outgrowth of a course given 10 years at M. I. T. Makes most useful branch of pure math accessible for practical work. Theoretical basis of D.E.'s; solution of ordinary D.E.'s and partial derivatives arising from heat flow, steady-state temperature of a plate, wave equations; analytic functions; convergence of Fourier Series. 400 problems on electricity, vibratory systems, other topics. Formerly "Differential Equations for Electrical Engineers." Index 41 illus. 307pp. 5⅜ x 8.
S601 Paperbound **$1.65**

DIFFERENTIAL EQUATIONS, F. R. Moulton. A detailed, rigorous exposition of all the non-elementary processes of solving ordinary differential equations. Several chapters devoted to the treatment of practical problems, especially those of a physical nature, which are far more advanced than problems usually given as illustrations. Includes analytic differential equations; variations of a parameter; integrals of differential equations; analytic implicit functions; problems of elliptic motion; sine-amplitude functions; deviation of formal bodies; Cauchy-Lipschitz process; linear differential equations with periodic coefficients; differential equations in infinitely many variations; much more. Historical notes. 10 figures. 222 problems. Index. xv + 395pp. 5⅜ x 8. S451 Paperbound **$2.00**

DIFFERENTIAL AND INTEGRAL EQUATIONS OF MECHANICS AND PHYSICS (DIE DIFFERENTIAL- UND INTEGRALGLEICHUNGEN DER MECHANIK UND PHYSIK), edited by P. Frank and R. von Mises. Most comprehensive and authoritative work on the mathematics of mathematical physics available today in the United States: the standard, definitive reference for teachers, physicists, engineers, and mathematicians—now published (in the original German) at a relatively inexpensive price for the first time! Every chapter in this 2,000-page set is by an expert in his field: Carathéodory, Courant, Frank, Mises, and a dozen others. Vol I, on mathematics, gives concise but complete coverages of advanced calculus, differential equations, integral equations, and potential, and partial differential equations. Index. xxiii + 916pp. Vol. II (physics): classical mechanics, optics, continuous mechanics, heat conduction and diffusion, the stationary and quasi-stationary electromagnetic field, electromagnetic oscillations, and wave mechanics. Index. xxiv + 1106pp. Two volume set. Each volume available separately. 5⅝ x 8⅜. S787 Vol I Clothbound **$7.50**
S788 Vol II Clothbound **$7.50**
The set **$15.00**

LECTURES ON CAUCHY'S PROBLEM, J. Hadamard. Based on lectures given at Columbia, Rome, this discusses work of Riemann, Kirchhoff, Volterra, and the author's own research on the hyperbolic case in linear partial differential equations. It extends spherical and cylindrical waves to apply to all (normal) hyperbolic equations. Partial contents: Cauchy's problem, fundamental formula, equations with odd number, with even number of independent variables; method of descent. 32 figures. Index. iii + 316pp. 5⅜ x 8. S105 Paperbound **$1.75**

THEORY OF DIFFERENTIAL EQUATIONS, A. R. Forsyth. Out of print for over a decade, the complete 6 volumes (now bound as 3) of this monumental work represent the most comprehensive treatment of differential equations ever written. Historical presentation includes in 2500 pages every substantial development. Vol. 1, 2: EXACT EQUATIONS, PFAFF'S PROBLEM; ORDINARY EQUATIONS, NOT LINEAR: methods of Grassmann, Clebsch, Lie, Darboux; Cauchy's theorem; branch points; etc. Vol. 3, 4: ORDINARY EQUATIONS, NOT LINEAR; ORDINARY LINEAR EQUATIONS: Zeta Fuchsian functions, general theorems on algebraic integrals, Brun's theorem, equations with uniform periodic coffiecients, etc. Vol. 4, 5: PARTIAL DIFFERENTIAL EQUATIONS: 2 existence-theorems, equations of theoretical dynamics, Laplace transformations, general transformation of equations of the 2nd order, much more. Indexes. Total of 2766pp. 5⅜ x 8. S576-7-8 Clothbound: the set **$15.00**

PARTIAL DIFFERENTIAL EQUATIONS OF MATHEMATICAL PHYSICS, A. G. Webster. A keystone work in the library of every mature physicist, engineer, researcher. Valuable sections on elasticity, compression theory, potential theory, theory of sound, heat conduction, wave propagation, vibration theory. Contents include: deduction of differential equations, vibrations, normal functions, Fourier's series, Cauchy's method, boundary problems, method of Riemann-Volterra. Spherical, cylindrical, ellipsoidal harmonics, applications, etc. 97 figures. vii + 440pp. 5⅜ x 8. S263 Paperbound **$2.00**

ORDINARY DIFFERENTIAL EQUATIONS, E. L. Ince. A most compendious analysis in real and complex domains. Existence and nature of solutions, continuous transformation groups, solutions in an infinite form, definite integrals, algebraic theory, Sturmian theory, boundary problems, existence theorems, 1st order, higher order, etc. "Deserves the highest praise, a notable addition to mathematical literature," BULLETIN, AM. MATH. SOC. Historical appendix. Bibliography. 18 figures. viii + 558pp. 5⅜ x 8. S349 Paperbound **$2.75**

INTRODUCTION TO NONLINEAR DIFFERENTIAL AND INTEGRAL EQUATIONS, Harold T. Davis. A thorough introduction to this important area, of increasing interest to mathematicians and scientists. First published by the United States Atomic Energy Commission, it includes chapters on the differential equation of the first order, the Riccati equation (as a bridge between linear and nonlinear equations), existence theorems, second order equations, elliptic integrals, elliptic functions, and theta functions, second order differential equations of polynomial class, continuous analytic continuation, the phase plane and its phenomena, nonlinear mechanics, the calculus of variations, etc. Appendices on Painlevé transcendents and Van der Pol and Volterra equations. Bibliography of 350 items. 137 problems. Index. xv + 566pp. 5⅜ x 8½. S971 Paperbound **$2.00**

THEORY OF FUNCTIONALS AND OF INTEGRAL AND INTEGRO-DIFFERENTIAL EQUATIONS, Vito Volterra. Unabridged republication of the only English translation. An exposition of the general theory of the functions depending on a continuous set of values of another function, based on the author's fundamental notion of the transition from a finite number of variables to a continually infinite number. Though dealing primarily with integral equations, much material on calculus of variations is included. The work makes no assumption of previous knowledge on the part of the reader. It begins with fundamental material and proceeds to Generalization of Analytic Functions, Integro-Differential Equations, Functional Derivative Equations, Applications, Other Directions of Theory of Functionals, etc. New introduction by G. C. Evans. Bibliography and criticism of Volterra's work by E. Whittaker. Bibliography. Index of authors cited. Index of subjects. xxxx + 226pp. 5⅜ x 8. S502 Paperbound **$1.75**

LINEAR INTEGRAL EQUATIONS, W. V. Lovitt. Systematic survey of general theory, with some application to differential equations, calculus of variations, problems of math, physics. Partial contents: integral equation of 2nd kind by successive substitutions; Fredholm's equation as ratio of 2 integral series in lambda, applications of the Fredholm theory, Hilbert-Schmidt theory of symmetric kernels, application, etc. Neumann, Dirichlet, vibratory problems. Index. ix + 253pp. 5⅜ x 8. S176 Paperbound **$1.60**

Foundations of mathematics

THE CONTINUUM AND OTHER TYPES OF SERIAL ORDER, E. V. Huntington. This famous book gives a systematic elementary account of the modern theory of the continuum as a type of serial order. Based on the Cantor-Dedekind ordinal theory, which requires no technical knowledge of higher mathematics, it offers an easily followed analysis of ordered classes, discrete and dense series, continuous series, Cantor's transfinite numbers. 2nd edition. Index. viii + 82pp. 5⅜ x 8. S130 Paperbound **$1.00**

CONTRIBUTIONS TO THE FOUNDING OF THE THEORY OF TRANSFINITE NUMBERS, Georg Cantor. These papers founded a new branch of mathematics. The famous articles of 1895-7 are translated, with an 82-page introduction by P. E. B. Jourdain dealing with Cantor, the background of his discoveries, their results, future possibilities. Bibliography. Index. Notes. ix + 211 pp. 5⅜ x 8. S45 Paperbound **$1.35**

ELEMENTARY MATHEMATICS FROM AN ADVANCED STANDPOINT, Felix Klein.

This classic text is an outgrowth of Klein's famous integration and survey course at Göttingen. Using one field of mathematics to interpret, adjust, illuminate another, it covers basic topics in each area, illustrating its discussion with extensive analysis. It is especially valuable in considering areas of modern mathematics. "Makes the reader feel the inspiration of . . . a great mathematician, inspiring teacher . . . with deep insight into the foundations and interrelations," BULLETIN, AMERICAN MATHEMATICAL SOCIETY.

Vol. 1. ARITHMETIC, ALGEBRA, ANALYSIS. Introducing the concept of function immediately, it enlivens abstract discussion with graphical and geometrically perceptual methods. Partial contents: natural numbers, extension of the notion of number, special properties, complex numbers. Real equations with real unknowns, complex quantities. Logarithmic, exponential functions, goniometric functions, infinitesimal calculus. Transcendence of e and pi, theory of assemblages. Index. 125 figures. ix + 274pp . 5⅜ x 8. S150 Paperbound **$1.85**

Vol. 2. GEOMETRY. A comprehensive view which accompanies the space perception inherent in geometry with analytic formulas which facilitate precise formulation. Partial contents: Simplest geometric manifolds: line segment, Grassmann determinant principles, classification of configurations of space, derivative manifolds. Geometric transformations: affine transformations, projective, higher point transformations, theory of the imaginary. Systematic discussion of geometry and its foundations. Indexes. 141 illustrations. ix + 214pp. 5⅜ x 8. S151 Paperbound **$1.75**

ESSAYS ON THE THEORY OF NUMBERS: 1. CONTINUITY AND IRRATIONAL NUMBERS; 2. THE NATURE AND MEANING OF NUMBERS, Richard Dedekind. The two most important essays on the logical foundations of the number system by the famous German mathematician. The first provides a purely arithmetic and perfectly rigorous foundation for irrational numbers and thereby a rigorous meaning to continuity in analysis. The second essay is an attempt to give a logical basis for transfinite numbers and properties of the natural numbers. Discusses the logical validity of mathematical induction. Authorized English translations by W. W. Deman of "Stetigkeit und irrationale Zahlen" and "Was sind und was sollen die Zahlen?" vii + 115pp. 5⅜ x 8. T1010 Paperbound **$1.00**

Geometry

THE FOUNDATIONS OF EUCLIDEAN GEOMETRY, H. G. Forder. The first rigorous account of Euclidean geometry, establishing propositions without recourse to empiricism, and without multiplying hypotheses. Corrects many traditional weaknesses of Euclidean proofs, and investigates the problems imposed on the axiom system by the discoveries of Bolyai and Lobachevsky. Some topics discussed are Classes and Relations; Axioms for Magnitudes; Congruence and Similarity; Algebra of Points; Hessenberg's Theorem; Continuity; Existence of Parallels; Reflections; Rotations; Isometries; etc. Invaluable for the light it throws on foundations of math. Lists: Axioms employed, Symbols, Constructions. 295pp. 5⅜ x 8. S481 Paperbound **$2.00**

ADVANCED EUCLIDEAN GEOMETRY, R. A. Johnson. For years the standard textbook on advanced Euclidean geometry, requires only high school geometry and trigonometry. Explores in unusual detail and gives proofs of hundreds of relatively recent theorems and corollaries, many formerly available only in widely scattered journals. Covers tangent circles, the theorem of Miquel, symmedian point, pedal triangles and circles, the Brocard configuration, and much more. Formerly "Modern Geometry." Index. 107 diagrams. xiii + 319pp. 5⅜ x 8. S669 Paperbound **$1.65**

HIGHER GEOMETRY: AN INTRODUCTION TO ADVANCED METHODS IN ANALYTIC GEOMETRY, F. S. Woods. Exceptionally thorough study of concepts and methods of advanced algebraic geometry (as distinguished from differential geometry). Exhaustive treatment of 1-, 2-, 3-, and 4-dimensional coordinate systems, leading to n-dimensional geometry in an abstract sense. Covers projectivity, tetracyclical coordinates, contact transformation, pentaspherical coordinates, much more. Based on M.I.T. lectures, requires sound preparation in analytic geometry and some knowledge of determinants. Index. Over 350 exercises. References. 60 figures. x + 423pp. 5⅜ x 8. S737 Paperbound **$2.00**

CONTEMPORARY GEOMETRY, André Delachet. Translated by Howard G. Bergmann. The recent developments in geometry covered in uncomplicated fashion. Clear discussions of modern thinking about the theory of groups, the concept of abstract geometry, projective geometry, algebraic geometry, vector spaces, new kinds of metric spaces, developments in differential geometry, etc. A large part of the book is devoted to problems, developments, and applications of topology. For advanced undergraduates and graduate students as well as mathematicians in other fields who want a brief introduction to current work in geometry. 39 figures. Index. xix + 94pp. 5⅜ x 8½. S988 Paperbound **$1.00**

ELEMENTS OF PROJECTIVE GEOMETRY, L. Cremona. Outstanding complete treatment of projective geometry by one of the foremost 19th century geometers. Detailed proofs of all fundamental principles, stress placed on the constructive aspects. Covers homology, law of duality, anharmonic ratios, theorems of Pascal and Brianchon, foci, polar reciprocal figures, etc. Only ordinary geometry necessary to understand this honored classic. Index. Over 150 fully worked out examples and problems. 252 diagrams. xx + 302pp. 5⅜ x 8. S668 Paperbound **$1.75**

AN INTRODUCTION TO PROJECTIVE GEOMETRY, R. M. Winger. One of the best introductory texts to an important area in modern mathematics. Contains full development of elementary concepts often omitted in other books. Employing the analytic method to capitalize on the student's collegiate training in algebra, analytic geometry and calculus, the author deals with such topics as Essential Constants, Duality, The Line at Infinity, Projective Properties and Double Ratio, Projective Coordinates, The Conic, Collineations and Involutions in One Dimension, Binary Forms, Algebraic Invariants, Analytic Treatment of the Conic, Collineations in the Plane, Cubic Involutions and the Rational Cubic Curve, and a clear discussion of Non-Euclidean Geometry. For senior-college students and graduates. "An excellent textbook . . . very clearly written . . . propositions stated concisely," A. Emch, Am. Math. Monthly. Corrected reprinting. 928 problems. Index. 116 figures. xii + 443pp. 5⅜ x 8. S949 Paperbound **$2.00**

ALGEBRAIC CURVES, Robert J. Walker, Professor of Mathematics, Cornell University. Fine introduction to algebraic geometry. Presents some of the recently developed algebraic methods of handling problems in algebraic geometry, shows how these methods are related to the older analytic and geometric problems, and applies them to those same geometric problems. Limited to the theory of curves, concentrating on birational transformations. Contents: Algebraic Preliminaries, Projective Spaces, Plane Algebraic Curves, Formal Power Series, Transformations of a Curve, Linear Series. 25 illustrations. Numerous exercises at ends of sections. Index. x + 201pp. 5⅜ x 8½. S336 Paperbound **$1.60**

THE ADVANCED GEOMETRY OF PLANE CURVES AND THEIR APPLICATIONS, C. Zwikker. An unusual study of many important curves, their geometrical properties and their applications, including discussions of many less well-known curves not often treated in textbooks on synthetic and analytic Euclidean geometry. Includes both algebraic and transcendental curves such as the conic sections, kinked curves, spirals, lemniscates, cycloids, etc. and curves generated as involutes, evolutes, anticaustics, pedals, envelopes and orthogonal trajectories. Dr. Zwikker represents the points of the curves by complex numbers instead of two real Cartesian coordinates, allowing direct and even elegant proofs. Formerly: "Advanced Plane Geometry." 273 figures. xii + 299pp. 5⅜ x 8½. S1078 Paperbound **$2.00**

A TREATISE ON THE DIFFERENTIAL GEOMETRY OF CURVES AND SURFACES, L. P. Eisenhart. Introductory treatise especially for the graduate student, for years a highly successful textbook. More detailed and concrete in approach than most more recent books. Covers space curves, osculating planes, moving axes, Gauss' method, the moving trihedral, geodesics, conformal representation, etc. Last section deals with deformation of surfaces, rectilinear congruences, cyclic systems, etc. Index. 683 problems. 30 diagrams. xii + 474pp. 5⅜ x 8. S667 Paperbound **$2.75**

A TREATISE ON ALGEBRAIC PLANE CURVES, J. L. Coolidge. Unabridged reprinting of one of few full coverages in English, offering detailed introduction to theory of algebraic plane curves and their relations to geometry and analysis. Treats topological properties, Riemann-Roch theorem, all aspects of wide variety of curves including real, covariant, polar, containing series of a given sort, elliptic, polygonal, rational, the pencil, two parameter nets, etc. This volume will enable the reader to appreciate the symbolic notation of Aronhold and Clebsch. Bibliography. Index. 17 illustrations. xxiv + 513pp. 5⅜ x 8. S543 Paperbound **$2.75**

AN INTRODUCTION TO THE GEOMETRY OF N DIMENSIONS, D. M. Y. Sommerville. An introduction presupposing no prior knowledge of the field, the only book in English devoted exclusively to higher dimensional geometry. Discusses fundamental ideas of incidence, parallelism, perpendicularity, angles between linear space; enumerative geometry; analytical geometry from projective and metric points of view; polytopes; elementary ideas in analysis situs; content of hyper-spacial figures. Bibliography. Index. 60 diagrams. 196pp. 5⅜ x 8. S494 Paperbound **$1.50**

GEOMETRY OF FOUR DIMENSIONS, H. P. Manning. Unique in English as a clear, concise introduction. Treatment is synthetic, and mostly Euclidean, although in hyperplanes and hyperspheres at infinity, non-Euclidean geometry is used. Historical introduction. Foundations of 4-dimensional geometry. Perpendicularity, simple angles. Angles of planes, higher order. Symmetry, order, motion; hyperpyramids, hypercones, hyperspheres; figures with parallel elements; volume, hypervolume in space; regular polyhedroids. Glossary. 78 figures. ix + 348pp. 5⅜ x 8. S182 Paperbound **$2.00**

CONVEX FIGURES AND POLYHEDRA, L. A. Lyusternik. An excellent elementary discussion by a leading Russian mathematician. Beginning with the basic concepts of convex figures and bodies and their supporting lines and planes, the author covers such matters as centrally symmetric convex figures, theorems of Euler, Cauchy, Steinitz and Alexandrov on convex polyhedra, linear systems of convex bodies, planar sections of convex bodies, the Brunn-Minkowski inequality and its consequences, and many other related topics. No more than a high school background in mathematics needed for complete understanding. First English translation by T. J. Smith. 182 illustrations. Index. x + 176pp. 5⅜ x 8½. S1021 Paperbound **$1.50**

NON-EUCLIDEAN GEOMETRY, Roberto Bonola. The standard coverage of non-Euclidean geometry. It examines from both a historical and mathematical point of view the geometries which have arisen from a study of Euclid's 5th postulate upon parallel lines. Also included are complete texts, translated, of Bolyai's SCIENCE OF ABSOLUTE SPACE. Lobachevsky's THEORY OF PARALLELS. 180 diagrams. 431pp. 5⅜ x 8. S27 Paperbound **$2.00**

ELEMENTS OF NON-EUCLIDEAN GEOMETRY, D. M. Y. Sommerville. Unique in proceeding step-by-step, in the manner of traditional geometry. Enables the student with only a good knowledge of high school algebra and geometry to grasp elementary hyperbolic, elliptic, analytic non-Euclidean geometries; space curvature and its philosophical implications; theory of radical axes; homothetic centres and systems of circles; parataxy and parallelism; absolute measure; Gauss' proof of the defect area theorem; geodesic representation; much more, all with exceptional clarity. 126 problems at chapter endings provide progressive practice and familiarity. 133 figures. Index. xvi + 274pp. 5⅜ x 8. S460 Paperbound **$1.50**

INTRODUCTORY NON-EUCLIDEAN GEOMETRY, H. P. Manning. Sound elementary introduction to non-Euclidean geometry. The first two thirds (Pangeometry and the Hyperbolic Geometry) require a grasp of plane and solid geometry and trigonometry. The last sections (the Elliptic Geometry and Analytic Non-Euclidean Geometry) necessitate also basic college calculus for understanding the text. The book does not propose to investigate the foundations of geometry, but rather begins with the theorems common to Euclidean and non-Euclidean geometry and then takes up the specific differences between them. A simple and direct account of the bases of this important branch of mathematics for teachers and students. 94 figures. vii + 95pp. 5⅜ x 8. S310 Paperbound **$1.00**

ELEMENTARY CONCEPTS OF TOPOLOGY, P. Alexandroff. First English translation of the famous brief introduction to topology for the beginner or for the mathematician not undertaking extensive study. This unusually useful intuitive approach deals primarily with the concepts of complex, cycle, and homology, and is wholly consistent with current investigations. Ranges from basic concepts of set-theoretic topology to the concept of Betti groups. "Glowing example of harmony between intuition and thought," David Hilbert. Translated by A. E. Farley. Introduction by D. Hilbert. Index. 25 figures. 73pp. 5⅜ x 8. S747 Paperbound **$1.00**

Number theory

INTRODUCTION TO THE THEORY OF NUMBERS, L. E. Dickson. Thorough, comprehensive approach with adequate coverage of classical literature, an introductory volume beginners can follow. Chapters on divisibility, congruences, quadratic residues & reciprocity, Diophantine equations, etc. Full treatment of binary quadratic forms without usual restriction to integral coefficients. Covers infinitude of primes, least residues, Fermat's theorem, Euler's phi function, Legendre's symbol, Gauss's lemma, automorphs, reduced forms, recent theorems of Thue & Siegel, many more. Much material not readily available elsewhere. 239 problems. Index. I figure. viii + 183pp. 5⅜ x 8. S342 Paperbound **$1.65**

ELEMENTS OF NUMBER THEORY, I. M. Vinogradov. Detailed 1st course for persons without advanced mathematics; 95% of this book can be understood by readers who have gone no farther than high school algebra. Partial contents: divisibility theory, important number theoretical functions, congruences, primitive roots and indices, etc. Solutions to both problems and exercises. Tables of primes, indices, etc. Covers almost every essential formula in elementary number theory! Translated from Russian. 233 problems, 104 exercises. viii + 227pp. 5⅜ x 8. S259 Paperbound **$1.60**

THEORY OF NUMBERS and DIOPHANTINE ANALYSIS, R. D. Carmichael. These two complete works in one volume form one of the most lucid introductions to number theory, requiring only a firm foundation in high school mathematics. "Theory of Numbers," partial contents: Eratosthenes' sieve, Euclid's fundamental theorem, G.C.F. and L.C.M. of two or more integers, linear congruences, etc "Diophantine Analysis": rational triangles, Pythagorean triangles, equations of third, fourth, higher degrees, method of functional equations, much more. "Theory of Numbers": 76 problems. Index. 94pp. "Diophantine Analysis": 222 problems. Index. 118pp. 5⅜ x 8. S529 Paperbound **$1.35**

Numerical analysis, tables

MATHEMATICAL TABLES AND FORMULAS, Compiled by Robert D. Carmichael and Edwin R. Smith. Valuable collection for students, etc. Contains all tables necessary in college algebra and trigonometry, such as five-place common logarithms, logarithmic sines and tangents of small angles, logarithmic trigonometric functions, natural trigonometric functions, four-place antilogarithms, tables for changing from sexagesimal to circular and from circular to sexagesimal measure of angles, etc. Also many tables and formulas not ordinarily accessible, including powers, roots, and reciprocals, exponential and hyperbolic functions, ten-place logarithms of prime numbers, and formulas and theorems from analytical and elementary geometry and from calculus. Explanatory introduction. viii + 269pp. 5⅜ x 8½. S111 Paperbound **$1.00**

MATHEMATICAL TABLES, H. B. Dwight. Unique for its coverage in one volume of almost every function of importance in applied mathematics, engineering, and the physical sciences. Three extremely fine tables of the three trig functions and their inverse functions to thousandths of radians; natural and common logarithms; squares, cubes; hyperbolic functions and the inverse hyperbolic functions; $(a^2 + b^2)$ exp. ½a; complete elliptic integrals of the 1st and 2nd kind; sine and cosine integrals; exponential integrals Ei(x) and Ei(—x); binomial coefficients; factorials to 250; surface zonal harmonics and first derivatives; Bernoulli and Euler numbers and their logs to base of 10; Gamma function; normal probability integral; over 60 pages of Bessel functions; the Riemann Zeta function. Each table with formulae generally used, sources of more extensive tables, interpolation data, etc. Over half have columns of differences, to facilitate interpolation. Introduction. Index. viii + 231pp. 5⅜ x 8. S445 Paperbound **$1.75**

TABLES OF FUNCTIONS WITH FORMULAE AND CURVES, E. Jahnke & F. Emde. The world's most comprehensive 1-volume English-text collection of tables, formulae, curves of transcendent functions. 4th corrected edition, new 76-page section giving tables, formulae for elementary functions—not in other English editions. Partial contents: sine, cosine, logarithmic integral; factorial function; error integral; theta functions; elliptic integrals, functions; Legendre, Bessel, Riemann, Mathieu, hypergeometric functions, etc. Supplementary books. Bibliography. Indexed. "Out of the way functions for which we know no other source," SCIENTIFIC COMPUTING SERVICE, Ltd. 212 figures. 400pp. 5⅜ x 8. S133 Paperbound **$2.00**

Probability theory and information theory

AN ELEMENTARY INTRODUCTION TO THE THEORY OF PROBABILITY, B. V. Gnedenko and A. Ya. Khinchin. Translated by Leo F. Boron. A clear, compact introduction designed to equip the reader with a fundamental grasp of the theory of probability. It is thorough and authoritative within its purposely restricted range, yet the layman with a background in elementary mathematics will be able to follow it without difficulty. Covers such topics as the processes involved in the calculation of probabilities, conditional probabilities and the multiplication rule, Bayes's formula, Bernoulli's scheme and theorem, random variables and distribution laws, and dispersion and mean deviations. New translation of fifth (revised) Russian edition (1960)—the only translation checked and corrected by Gnedenko. New preface for Dover edition by B. V. Gnedenko. Index. Bibliography. Appendix: Table of values of function $\phi(a)$. xii + 130pp. 5⅜ x 8½.
T155 Paperbound **$1.50**

AN INTRODUCTION TO MATHEMATICAL PROBABILITY, Julian Lowell Coolidge. A thorough introduction which presents the mathematical foundation of the theory of probability. A substantial body of material, yet can be understood with a knowledge of only elementary calculus. Contains: The Scope and Meaning of Mathematical Probability; Elementary Principles of Probability; Bernoulli's Theorem; Mean Value and Dispersion; Geometrical Probability; Probability of Causes; Errors of Observation; Errors in Many Variables; Indirect Observations; The Statistical Theory of Gases; and The Principles of Life Insurance. Six pages of logarithm tables. 4 diagrams. Subject and author indices. xii + 214pp. 5⅜ x 8½.
S258 Paperbound **$1.35**

A GUIDE TO OPERATIONS RESEARCH, W. E. Duckworth. A brief nontechnical exposition of techniques and theories of operational research. A good introduction for the layman; also can provide the initiate with new understandings. No mathematical training needed, yet not an oversimplification. Covers game theory, mathematical analysis, information theory, linear programming, cybernetics, decision theory, etc. Also includes a discussion of the actual organization of an operational research program and an account of the uses of such programs in the oil, chemical, paper, and metallurgical industries, etc. Bibliographies at chapter ends. Appendices. 36 figures. 145pp. 5¼ x 8½. T1129 Clothbound **$3.50**

MATHEMATICAL FOUNDATIONS OF INFORMATION THEORY, A. I. Khinchin. For the first time mathematicians, statisticians, physicists, cyberneticists, and communications engineers are offered a complete and exact introduction to this relatively new field. Entropy as a measure of a finite scheme, applications to coding theory, study of sources, channels and codes, detailed proofs of both Shannon theorems for any ergodic source and any stationary channel with finite memory, and much more are covered. Bibliography. vii + 120pp. 5⅜ x 8.
S434 Paperbound **$1.35**

SELECTED PAPERS ON NOISE AND STOCHASTIC PROCESS, edited by **Prof. Nelson Wax,** U. of Illinois. 6 basic papers for newcomers in the field, for those whose work involves noise characteristics. Chandrasekhar, Uhlenbeck & Ornstein, Uhlenbeck & Ming, Rice, Doob. Included is Kac's Chauvenet-Prize winning Random Walk. Extensive bibliography lists 200 articles, up through 1953. 21 figures. 337pp. 6⅛ x 9¼. S262 Paperbound **$2.50**

THEORY OF PROBABILITY, William Burnside. Synthesis, expansion of individual papers presents numerous problems in classical probability, offering many original views succinctly, effectively. Game theory, cards, selections from groups; geometrical probability in such areas as suppositions as to probability of position of point on a line, points on surface of sphere, etc. Includes methods of approximation, theory of errors, direct calculation of probabilities, etc. Index. 136pp. 5⅜ x 8. S567 Paperbound **$1.00**

Statistics

ELEMENTARY STATISTICS, WITH APPLICATIONS IN MEDICINE AND THE BIOLOGICAL SCIENCES, F. E. Croxton. A sound introduction to statistics for anyone in the physical sciences, assuming no prior acquaintance and requiring only a modest knowledge of math. All basic formulas carefully explained and illustrated; all necessary reference tables included. From basic terms and concepts, the study proceeds to frequency distribution, linear, non-linear, and multiple correlation, skewness, kurtosis, etc. A large section deals with reliability and significance of statistical methods. Containing concrete examples from medicine and biology, this book will prove unusually helpful to workers in those fields who increasingly must evaluate, check, and interpret statistics. Formerly titled "Elementary Statistics with Applications in Medicine." 101 charts. 57 tables. 14 appendices. Index. iv + 376pp. 5⅜ x 8.
S506 Paperbound **$2.00**

ANALYSIS & DESIGN OF EXPERIMENTS, H. B. Mann. Offers a method for grasping the analysis of variance and variance design within a short time. Partial contents: Chi-square distribution and analysis of variance distribution, matrices, quadratic forms, likelihood ration tests and tests of linear hypotheses, power of analysis, Galois fields, non-orthogonal data, interblock estimates, etc. 15pp. of useful tables. x + 195pp. 5 x 7⅜. S180 Paperbound **$1.45**

METHODS OF STATISTICS, L. H. C. Tippett. A classic in its field, this unusually complete systematic introduction to statistical methods begins at beginner's level and progresses to advanced levels for experimenters and poll-takers in all fields of statistical research. Supplies fundamental knowledge of virtually all elementary methods in use today by sociologists, psychologists, biologists, engineers, mathematicians, etc. Explains logical and mathematical basis of each method described, with examples for each section. Covers frequency distributions and measures, inference from random samples, errors in large samples, simple analysis of variance, multiple and partial regression and correlation, etc. 4th revised (1952) edition. 16 charts. 5 significance tables. 152-item bibliography. 96 tables. 22 figures. 395pp. 6 x 9.
S228 Clothbound **$7.50**

STATISTICS MANUAL, E. L. Crow, F. A. Davis, M. W. Maxfield. Comprehensive collection of classical, modern statistics methods, prepared under auspices of U. S. Naval Ordnance Test Station, China Lake, Calif. Many examples from ordnance will be valuable to workers in all fields. Emphasis is on use, with information on fiducial limits, sign tests, Chi-square runs, sensitivity, quality control, much more. "Well written . . . excellent reference work," Operations Research. Corrected edition of NAVORD Report 3360 NOTS 948. Introduction. Appendix of 32 tables, charts. Index. Bibliography. 95 illustrations. 306pp. 5⅜ x 8.
S599 Paperbound **$1.75**

Symbolic logic

AN INTRODUCTION TO SYMBOLIC LOGIC, Susanne K. Langer. Probably the clearest book ever written on symbolic logic for the philosopher, general scientist and layman. It will be particularly appreciated by those who have been rebuffed by other introductory works because of insufficient mathematical training. No special knowledge of mathematics is required. Starting with the simplest symbols and conventions, you are led to a remarkable grasp of the Boole-Schroeder and Russell-Whitehead systems clearly and quickly. PARTIAL CONTENTS: Study of forms, Essentials of logical structure, Generalization, Classes, The deductive system of classes, The algebra of logic, Abstraction of interpretation, Calculus of propositions, Assumptions of PRINCIPIA MATHEMATICA, Logistics, Logic of the syllogism, Proofs of theorems. "One of the clearest and simplest introductions to a subject which is very much alive. The style is easy, symbolism is introduced gradually, and the intelligent non-mathematician should have no difficulty in following the argument," MATHEMATICS GAZETTE. Revised, expanded second edition. Truth-value tables. 368pp. 5⅜ x 8.
S164 Paperbound **$1.85**

A SURVEY OF SYMBOLIC LOGIC: THE CLASSIC ALGEBRA OF LOGIC, C. I. Lewis. Classic survey of the field, comprehensive and thorough. Indicates content of major systems, alternative methods of procedure, and relation of these to the Boole-Schroeder algebra and to one another. Contains historical summary, as well as full proofs and applications of the classic, or Boole-Schroeder, algebra of logic. Discusses diagrams for the logical relations of classes, the two-valued algebra, propositional functions of two or more variables, etc. Chapters 5 and 6 of the original edition, which contained material not directly pertinent, have been omitted in this edition at the author's request. Appendix. Bibliography. Index. viii + 352pp. 5⅝ x 8⅜.
S643 Paperbound **$2.00**

INTRODUCTION TO SYMBOLIC LOGIC AND ITS APPLICATIONS, R. Carnap. One of the clearest, most comprehensive, and rigorous introductions to modern symbolic logic by perhaps its greatest living master. Symbolic languages are analyzed and one constructed. Applications to math (symbolic representation of axiom systems for set theory, natural numbers, real numbers, topology, Dedekind and Cantor explanations of continuity), physics (the general analysis of concepts of determination, causality, space-time-topology, based on Einstein), biology (symbolic representation of an axiom system for basic concepts). "A masterpiece," Zentralblatt für Mathematik und ihre Grenzgebiete. Over 300 exercises. 5 figures. Bibliography. Index. xvi + 241pp. 5⅜ x 8.
S453 Paperbound **$1.85**
Clothbound **$4.00**

SYMBOLIC LOGIC, C. I. Lewis, C. H. Langford. Probably the most cited book in symbolic logic, this is one of the fullest treatments of paradoxes. A wide coverage of the entire field of symbolic logic, plus considerable material that has not appeared elsewhere. Basic to the entire volume is the distinction between the logic of extensions and of intensions. Considerable emphasis is placed on converse substitution, while the matrix system presents the supposition of a variety of non-Aristotelian logics. It has especially valuable sections on strict limitations, existence of terms, 2-valued algebra and its extension to propositional functions, truth value systems, the matrix method, implication and deductibility, general theory of propositions, propositions of ordinary discourse, and similar topics. "Authoritative, most valuable," TIMES, London. Bibliography. 506pp. 5⅜ x 8. S170 Paperbound **$2.35**

THE ELEMENTS OF MATHEMATICAL LOGIC, Paul Rosenbloom. First publication in any language. This book is intended for readers who are mature mathematically, but have no previous training in symbolic logic. It does not limit itself to a single system, but covers the field as a whole. It is a development of lectures given at Lund University, Sweden, in 1948. Partial contents: Logic of classes, fundamental theorems, Boolean algebra, logic of propositions, logic of propositional functions, expressive languages, combinatory logics, development of mathematics within an object language, paradoxes, theorems of Post and Goedel, Church's theorem, and similar topics. iv + 214pp. 5⅜ x 8. S227 Paperbound **$1.45**

BOOKS EXPLAINING SCIENCE AND MATHEMATICS

General

WHAT IS SCIENCE?, Norman Campbell. This excellent introduction explains scientific method, role of mathematics, types of scientific laws. Contents: 2 aspects of science, science & nature, laws of science, discovery of laws, explanation of laws, measurement & numerical laws, applications of science. 192pp. 5⅜ x 8. S43 Paperbound **$1.25**

THE COMMON SENSE OF THE EXACT SCIENCES, W. K. Clifford. Introduction by James Newman, edited by Karl Pearson. For 70 years this has been a guide to classical scientific and mathematical thought. Explains with unusual clarity basic concepts, such as extension of meaning of symbols, characteristics of surface boundaries, properties of plane figures, vectors, Cartesian method of determining position, etc. Long preface by Bertrand Russell. Bibliography of Clifford. Corrected, 130 diagrams redrawn. 249pp. 5⅜ x 8. T61 Paperbound **$1.60**

SCIENCE THEORY AND MAN, Erwin Schrödinger. This is a complete and unabridged reissue of SCIENCE AND THE HUMAN TEMPERAMENT plus an additional essay: "What is an Elementary Particle?" Nobel laureate Schrödinger discusses such topics as nature of scientific method, the nature of science, chance and determinism, science and society, conceptual models for physical entities, elementary particles and wave mechanics. Presentation is popular and may be followed by most people with little or no scientific training. "Fine practical preparation for a time when laws of nature, human institutions . . . are undergoing a critical examination without parallel," Waldemar Kaempffert, N. Y. TIMES. 192pp. 5⅜ x 8. T428 Paperbound **$1.35**

FADS AND FALLACIES IN THE NAME OF SCIENCE, Martin Gardner. Examines various cults, quack systems, frauds, delusions which at various times have masqueraded as science. Accounts of hollow-earth fanatics like Symmes; Velikovsky and wandering planets; Hoerbiger; Bellamy and the theory of multiple moons; Charles Fort; dowsing, pseudoscientific methods for finding water, ores, oil. Sections on naturopathy, iridiagnosis, zone therapy, food fads, etc. Analytical accounts of Wilhelm Reich and orgone sex energy; L. Ron Hubbard and Dianetics; A. Korzybski and General Semantics; many others. Brought up to date to include Bridey Murphy, others. Not just a collection of anecdotes, but a fair, reasoned appraisal of eccentric theory. Formerly titled IN THE NAME OF SCIENCE. Preface. Index. x + 384pp. 5⅜ x 8. T394 Paperbound **$1.50**

A DOVER SCIENCE SAMPLER, edited by George Barkin. 64-page book, sturdily bound, containing excerpts from over 20 Dover books, explaining science. Edwin Hubble, George Sarton, Ernst Mach, A. d'Abro, Galileo, Newton, others, discussing island universes, scientific truth, biological phenomena, stability in bridges, etc. Copies limited; no more than 1 to a customer, FREE

POPULAR SCIENTIFIC LECTURES, Hermann von Helmholtz. Helmholtz was a superb expositor as well as a scientist of genius in many areas. The seven essays in this volume are models of clarity, and even today they rank among the best general descriptions of their subjects ever written. "The Physiological Causes of Harmony in Music" was the first significant physiological explanation of musical consonance and dissonance. Two essays, "On the Interaction of Natural Forces" and "On the Conservation of Force," were of great importance in the history of science, for they firmly established the principle of the conservation of energy. Other lectures include "On the Relation of Optics to Painting," "On Recent Progress in the Theory of Vision," "On Goethe's Scientific Researches," and "On the Origin and Significance of Geometrical Axioms." Selected and edited with an introduction by Professor Morris Kline. xii + 286pp. 5⅜ x 8½. T799 Paperbound **$1.45**

BOOKS EXPLAINING SCIENCE AND MATHEMATICS

Physics

CONCERNING THE NATURE OF THINGS, Sir William Bragg. Christmas lectures delivered at the Royal Society by Nobel laureate. Why a spinning ball travels in a curved track; how uranium is transmuted to lead, etc. Partial contents: atoms, gases, liquids, crystals, metals, etc. No scientific background needed; wonderful for intelligent child. 32pp. of photos, 57 figures. xii + 232pp. 5⅜ x 8. T31 Paperbound **$1.50**

THE RESTLESS UNIVERSE, Max Born. New enlarged version of this remarkably readable account by a Nobel laureate. Moving from sub-atomic particles to universe, the author explains in very simple terms the latest theories of wave mechanics. Partial contents: air and its relatives, electrons & ions, waves & particles, electronic structure of the atom, nuclear physics. Nearly 1000 illustrations, including 7 animated sequences. 325pp. 6 x 9. T412 Paperbound **$2.00**

THE STRANGE STORY OF THE QUANTUM, AN ACCOUNT FOR THE GENERAL READER OF THE GROWTH OF IDEAS UNDERLYING OUR PRESENT ATOMIC KNOWLEDGE, B. Hoffmann. Presents lucidly and expertly, with barest amount of mathematics, the problems and theories which led to modern quantum physics. Dr. Hoffmann begins with the closing years of the 19th century, when certain trifling discrepancies were noticed, and with illuminating analogies and examples takes you through the brilliant concepts of Planck, Einstein, Pauli, de Broglie, Bohr, Schroedinger, Heisenberg, Dirac, Sommerfeld, Feynman, etc. This edition includes a new, long postscript carrying the story through 1958. "Of the books attempting an account of the history and contents of our modern atomic physics which have come to my attention, this is the best," H. Margenau, Yale University, in "American Journal of Physics." 32 tables and line illustrations. Index. 275pp. 5⅜ x 8. T518 Paperbound **$1.50**

THE EVOLUTION OF SCIENTIFIC THOUGHT FROM NEWTON TO EINSTEIN, A. d'Abro. Einstein's special and general theories of relativity, with their historical implications, are analyzed in non-technical terms. Excellent accounts of the contributions of Newton, Riemann, Weyl, Planck, Eddington, Maxwell, Lorentz and others are treated in terms of space and time, equations of electromagnetics, finiteness of the universe, methodology of science. 21 diagrams. 482pp. 5⅜ x 8. T2 Paperound **$2.25**

THE RISE OF THE NEW PHYSICS, A. d'Abro. A half-million word exposition, formerly titled THE DECLINE OF MECHANISM, for readers not versed in higher mathematics. The only thorough explanation, in everyday language, of the central core of modern mathematical physical theory, treating both classical and modern theoretical physics, and presenting in terms almost anyone can understand the equivalent of 5 years of study of mathematical physics. Scientifically impeccable coverage of mathematical-physical thought from the Newtonian system up through the electronic theories of Dirac and Heisenberg and Fermi's statistics. Combines both history and exposition; provides a broad yet unified and detailed view, with constant comparison of classical and modern views on phenomena and theories. "A must for anyone doing serious study in the physical sciences," JOURNAL OF THE FRANKLIN INSTITUTE. "Extraordinary faculty . . . to explain ideas and theories of theoretical physics in the language of daily life," ISIS. First part of set covers philosophy of science, drawing upon the practice of Newton, Maxwell, Poincaré, Einstein, others, discussing modes of thought, experiment, interpretations of causality, etc. In the second part, 100 pages explain grammar and vocabulary of mathematics, with discussions of functions, groups, series, Fourier series, etc. The remainder is devoted to concrete, detailed coverage of both classical and quantum physics, explaining such topics as analytic mechanics, Hamilton's principle, wave theory of light, electromagnetic waves, groups of transformations, thermodynamics, phase rule, Brownian movement, kinetics, special relativity, Planck's original quantum theory, Bohr's atom, Zeeman effect, Broglie's wave mechanics, Heisenberg's uncertainty, Eigen-values, matrices, scores of other important topics. Discoveries and theories are covered for such men as Alembert, Born, Cantor, Debye, Euler, Foucault, Galois, Gauss, Hadamard, Kelvin, Kepler, Laplace, Maxwell, Pauli, Rayleigh, Volterra, Weyl, Young, more than 180 others. Indexed. 97 illustrations. ix + 982pp. 5⅜ x 8. T3 Volume 1, Paperbound **$2.25**
T4 Volume 2, Paperbound **$2.25**

SPINNING TOPS AND GYROSCOPIC MOTION, John Perry. Well-known classic of science still unsurpassed for lucid, accurate, delightful exposition. How quasi-rigidity is induced in flexible and fluid bodies by rapid motions; why gyrostat falls, top rises; nature and effect on climatic conditions of earth's precessional movement; effect of internal fluidity on rotating bodies, etc. Appendixes describe practical uses to which gyroscopes have been put in ships, compasses, monorail transportation. 62 figures. 128pp. 5⅜ x 8. T416 Paperbound **$1.00**

THE UNIVERSE OF LIGHT, Sir William Bragg. No scientific training needed to read Nobel Prize winner's expansion of his Royal Institute Christmas Lectures. Insight into nature of light, methods and philosophy of science. Explains lenses, reflection, color, resonance, polarization, x-rays, the spectrum, Newton's work with prisms, Huygens' with polarization, Crookes' with cathode ray, etc. Leads into clear statement of 2 major historical theories of light, corpuscle and wave. Dozens of experiments you can do. 199 illus., including 2 full-page color plates. 293pp. 5⅜ x 8. S538 Paperbound **$1.85**

THE STORY OF X-RAYS FROM RÖNTGEN TO ISOTOPES, A. R. Bleich. Non-technical history of x-rays, their scientific explanation, their applications in medicine, industry, research, and art, and their effect on the individual and his descendants. Includes amusing early reactions to Röntgen's discovery, cancer therapy, detections of art and stamp forgeries, potential risks to patient and operator, etc. Illustrations show x-rays of flower structure, the gall bladder, gears with hidden defects, etc. Original Dover publication. Glossary. Bibliography. Index. 55 photos and figures. xiv + 186pp. 5⅜ x 8. T662 Paperbound **$1.35**

ELECTRONS, ATOMS, METALS AND ALLOYS, Wm. Hume-Rothery. An introductory-level explanation of the application of the electronic theory to the structure and properties of metals and alloys, taking into account the new theoretical work done by mathematical physicists. Material presented in dialogue-form between an "Old Metallurgist" and a "Young Scientist." Their discussion falls into 4 main parts: the nature of an atom, the nature of a metal, the nature of an alloy, and the structure of the nucleus. They cover such topics as the hydrogen atom, electron waves, wave mechanics, Brillouin zones, co-valent bonds, radioactivity and natural disintegration, fundamental particles, structure and fission of the nucleus,etc. Revised, enlarged edition. 177 illustrations. Subject and name indexes. 407pp. 5⅜ x 8½. S1046 Paperbound **$2.25**

FROM EUCLID TO EDDINGTON: A STUDY OF THE CONCEPTIONS OF THE EXTERNAL WORLD, Sir Edmund Whittaker. A foremost British scientist traces the development of theories of natural philosophy from the western rediscovery of Euclid to Eddington, Einstein, Dirac, etc. The inadequacy of classical physics is contrasted with present day attempts to understand the physical world through relativity, non-Euclidean geometry, space curvature, wave mechanics, etc. 5 major divisions of examination: Space; Time and Movement; the Concepts of Classical Physics; the Concepts of Quantum Mechanics; the Eddington Universe. 212pp. 5⅜ x 8. T491 Paperbound **$1.35**

PHYSICS, THE PIONEER SCIENCE, L. W. Taylor. First thorough text to place all important physical phenomena in cultural-historical framework; remains best work of its kind. Exposition of physical laws, theories developed chronologically, with great historical, illustrative experiments diagrammed, described, worked out mathematically. Excellent physics text for self-study as well as class work. Vol. 1: Heat, Sound: motion, acceleration, gravitation, conservation of energy, heat engines, rotation, heat, mechanical energy, etc. 211 illus. 407pp. 5⅜ x 8. Vol. 2: Light, Electricity: images, lenses, prisms, magnetism, Ohm's law, dynamos, telegraph, quantum theory, decline of mechanical view of nature, etc. Bibliography. 13 table appendix. Index. 551 illus. 2 color plates. 508pp. 5⅜ x 8.

Vol. 1 S565 Paperbound **$2.00**
Vol. 2 S566 Paperbound **$2.00**
The set **$4.00**

A SURVEY OF PHYSICAL THEORY, Max Planck. One of the greatest scientists of all time, creator of the quantum revolution in physics, writes in non-technical terms of his own discoveries and those of other outstanding creators of modern physics. Planck wrote this book when science had just crossed the threshold of the new physics, and he communicates the excitement felt then as he discusses electromagnetic theories, statistical methods, evolution of the concept of light, a step-by-step description of how he developed his own momentous theory, and many more of the basic ideas behind modern physics. Formerly "A Survey of Physics." Bibliography. Index. 128pp. 5⅜ x 8. S650 Paperbound **$1.15**

THE ATOMIC NUCLEUS, M. Korsunsky. The only non-technical comprehensive account of the atomic nucleus in English. For college physics students, etc. Chapters cover: Radioactivity, the Nuclear Model of the Atom, the Mass of Atomic Nuclei, the Disintegration of Atomic Nuclei, the Discovery of the Positron, the Artificial Transformation of Atomic Nuclei, Artificial Radioactivity, Mesons, the Neutrino, the Structure of Atomic Nuclei and Forces Acting Between Nuclear Particles, Nuclear Fission, Chain Reaction, Peaceful Uses, Thermoculear Reactions. Slightly abridged edition. Translated by G. Yankovsky. 65 figures. Appendix includes 45 photographic illustrations. 413 pp. 5⅜ x 8. S1052 Paperbound **$2.00**

PRINCIPLES OF MECHANICS SIMPLY EXPLAINED, Morton Mott-Smith. Excellent, highly readable introduction to the theories and discoveries of classical physics. Ideal for the layman who desires a foundation which will enable him to understand and appreciate contemporary developments in the physical sciences. Discusses: Density, The Law of Gravitation, Mass and Weight, Action and Reaction, Kinetic and Potential Energy, The Law of Inertia, Effects of Acceleration, The Independence of Motions, Galileo and the New Science of Dynamics, Newton and the New Cosmos, The Conservation of Momentum, and other topics. Revised edition of "This Mechanical World." Illustrated by E. Kosa, Jr. Bibliography and Chronology. Index. xiv + 171pp. 5⅜ x 8½. T1067 Paperbound **$1.00**

THE CONCEPT OF ENERGY SIMPLY EXPLAINED, Morton Mott-Smith. Elementary, non-technical exposition which traces the story of man's conquest of energy, with particular emphasis on the developments during the nineteenth century and the first three decades of our own century. Discusses man's earlier efforts to harness energy, more recent experiments and discoveries relating to the steam engine, the engine indicator, the motive power of heat, the principle of excluded perpetual motion, the bases of the conservation of energy, the concept of entropy, the internal combustion engine, mechanical refrigeration, and many other related topics. Also much biographical material. Index. Bibliography. 33 illustrations. ix + 215pp. 5⅜ x 8½. T1071 Paperbound **$1.25**

HEAT AND ITS WORKINGS, Morton Mott-Smith. One of the best elementary introductions to the theory and attributes of heat, covering such matters as the laws governing the effect of heat on solids, liquids and gases, the methods by which heat is measured, the conversion of a substance from one form to another through heating and cooling, evaporation, the effects of pressure on boiling and freezing points, and the three ways in which heat is transmitted (conduction, convection, radiation). Also brief notes on major experiments and discoveries. Concise, but complete, it presents all the essential facts about the subject in readable style. Will give the layman and beginning student a first-rate background in this major topic in physics. Index. Bibliography. 50 illustrations. x + 165pp. 5⅜ x 8½. T978 Paperbound **$1.00**

THE STORY OF ATOMIC THEORY AND ATOMIC ENERGY, J. G. Feinberg. Wider range of facts on physical theory, cultural implications, than any other similar source. Completely non-technical. Begins with first atomic theory, 600 B.C., goes through A-bomb, developments to 1959. Avogadro, Rutherford, Bohr, Einstein, radioactive decay, binding energy, radiation danger, future benefits of nuclear power, dozens of other topics, told in lively, related, informal manner. Particular stress on European atomic research. "Deserves special mention . . . authoritative," Saturday Review. Formerly "The Atom Story." New chapter to 1959. Index. 34 illustrations. 251pp. 5⅜ x 8. T625 Paperbound **$1.60**

TEACH YOURSELF MECHANICS, P. Abbott. The lever, centre of gravity, parallelogram of force, friction, acceleration, Newton's laws of motion, machines, specific gravity, gas, liquid pressure, much more. 280 problems, solutions. Tables. 163 illus. 271pp. 6⅞ x 4¼. Clothbound **$2.00**

MATTER & MOTION, James Clerk Maxwell, This excellent exposition begins with simple particles and proceeds gradually to physical systems beyond complete analysis: motion, force, properties of centre of mass of material system, work, energy, gravitation, etc. Written with all Maxwell's original insights and clarity. Notes by E. Larmor. 17 diagrams. 178pp. 5⅜ x 8. S188 Paperbound **$1.35**

SOAP BUBBLES, THEIR COLOURS AND THE FORCES WHICH MOULD THEM, C. V. Boys. Only complete edition, half again as much material as any other. Includes Boys' hints on performing his experiments, sources of supply. Dozens of lucid experiments show complexities of liquid films, surface tension, etc. Best treatment ever written. Introduction. 83 illustrations. Color plate. 202pp. 5⅜ x 8. T542 Paperbound **95¢**

MATTER & LIGHT, THE NEW PHYSICS, L. de Broglie. Non-technical papers by a Nobel laureate explain electromagnetic theory, relativity, matter, light and radiation, wave mechanics, quantum physics, philosophy of science. Einstein, Planck, Bohr, others explained so easily that no mathematical training is needed for all but 2 of the 21 chapters. Unabridged. Index. 300pp. 5⅜ x 8. T35 Paperbound **$1.85**

SPACE AND TIME, Emile Borel. An entirely non-technical introduction to relativity, by world-renowned mathematician, Sorbonne professor. (Notes on basic mathematics are included separately.) This book has never been surpassed for insight, and extraordinary clarity of thought, as it presents scores of examples, analogies, arguments, illustrations, which explain such topics as: difficulties due to motion; gravitation a force of inertia; geodesic lines; wave-length and difference of phase; x-rays and crystal structure; the special theory of relativity; and much more. Indexes. 4 appendixes. 15 figures. xvi + 243pp. 5⅜ x 8. T592 Paperbound **$1.45**

BOOKS EXPLAINING SCIENCE AND MATHEMATICS

Astronomy

THE FRIENDLY STARS, Martha Evans Martin. This engaging survey of stellar lore and science is a well-known classic, which has introduced thousands to the fascinating world of stars and other celestial bodies. Descriptions of Capella, Sirius, Arcturus, Vega, Polaris, etc.—all the important stars, with informative discussions of rising and setting of stars, their number, names, brightness, distances, etc. in a non-technical, highly readable style. Also: double stars, constellations, clusters—concentrating on stars and formations visible to the naked eye. New edition, revised (1963) by D. H. Menzel, Director Harvard Observatory. 23 diagrams by Prof. Ching-Sung Yu. Foreword by D. H. Menzel and W. W. Morgan. 2 Star Charts. Index. xii + 147pp. 5⅜ x 8½. T1099 Paperbound **$1.00**

AN ELEMENTARY SURVEY OF CELESTIAL MECHANICS, Y. Ryabov. Elementary exposition of gravitational theory and celestial mechanics. Historical introduction and coverage of basic principles, including: the elliptic, the orbital plane, the 2- and 3-body problems, the discovery of Neptune, planetary rotation, the length of the day, the shapes of galaxies, satellites (detailed treatment of Sputnik I), etc. First American reprinting of successful Russian popular exposition. Elementary algebra and trigonometry helpful, but not necessary; presentation chiefly verbal. Appendix of theorem proofs. 58 figures. 165pp. 5⅜ x 8. T756 Paperbound **$1.25**

THE SKY AND ITS MYSTERIES, E. A. Beet. One of most lucid books on mysteries of universe; deals with astronomy from earliest observations to latest theories of expansion of universe, source of stellar energy, birth of planets, origin of moon craters, possibility of life on other planets. Discusses effects of sunspots on weather; distances, ages of several stars; master plan of universe; methods and tools of astronomers; much more. "Eminently readable book," London Times. Extensive bibliography. Over 50 diagrams. 12 full-page plates, fold-out star map. Introduction. Index. 5¼ x 7½. T627 Clothbound **$3.50**

THE REALM OF THE NEBULAE, E. Hubble. One of the great astronomers of our time records his formulation of the concept of "island universes," and its impact on astronomy. Such topics are covered as the velocity-distance relation; classification, nature, distances, general field of nebulae; cosmological theories; nebulae in the neighborhood of the Milky Way. 39 photos of nebulae, nebulae clusters, spectra of nebulae, and velocity distance relations shown by spectrum comparison. "One of the most progressive lines of astronomical research," The Times (London). New introduction by A. Sandage. 55 illustrations. Index. iv + 201pp. 5⅜ x 8. S455 Paperbound **$1.50**

OUT OF THE SKY, H. H. Nininger. A non-technical but comprehensive introduction to "meteoritics", the young science concerned with all aspects of the arrival of matter from outer space. Written by one of the world's experts on meteorites, this work shows how, despite difficulties of observation and sparseness of data, a considerable body of knowledge has arisen. It defines meteors and meteorites; studies fireball clusters and processions, meteorite composition, size, distribution, showers, explosions, origins, craters, and much more. A true connecting link between astronomy and geology. More than 175 photos, 22 other illustrations. References. Bibliography of author's publications on meteorites. Index. viii + 336pp. 5⅜ x 8.
T519 Paperbound **$1.85**

SATELLITES AND SCIENTIFIC RESEARCH, D. King-Hele. Non-technical account of the manmade satellites and the discoveries they have yielded up to the autumn of 1961. Brings together information hitherto published only in hard-to-get scientific journals. Includes the life history of a typical satellite, methods of tracking, new information on the shape of the earth, zones of radiation, etc. Over 60 diagrams and 6 photographs. Mathematical appendix. Bibliography of over 100 items. Index. xii + 180pp. 5⅜ x 8½.
T703 Paperbound **$2.00**

BOOKS EXPLAINING SCIENCE AND MATHEMATICS

Mathematics

CHANCE, LUCK AND STATISTICS: THE SCIENCE OF CHANCE, Horace C. Levinson. Theory of probability and science of statistics in simple, non-technical language. Part I deals with theory of probability, covering odd superstitions in regard to "luck," the meaning of betting odds, the law of mathematical expectation, gambling, and applications in poker, roulette, lotteries, dice, bridge, and other games of chance. Part II discusses the misuse of statistics, the concept of statistical probabilities, normal and skew frequency distributions, and statistics applied to various fields—birth rates, stock speculation, insurance rates, advertising, etc. "Presented in an easy humorous style which I consider the best kind of expository writing," Prof. A. C. Cohen, Industry Quality Control. Enlarged revised edition. Formerly titled "The Science of Chance." Preface and two new appendices by the author. Index. xiv + 365pp. 5⅜ x 8.
T1007 Paperbound **$1.85**

PROBABILITIES AND LIFE, Emile Borel. Translated by M. Baudin. Non-technical, highly readable introduction to the results of probability as applied to everyday situations. Partial contents: Fallacies About Probabilities Concerning Life After Death; Negligible Probabilities and the Probabilities of Everyday Life; Events of Small Probability; Application of Probabilities to Certain Problems of Heredity; Probabilities of Deaths, Diseases, and Accidents; On Poisson's Formula. Index. 3 Appendices of statistical studies and tables. vi + 87pp. 5⅜ x 8½.
T121 Paperbound **$1.00**

GREAT IDEAS OF MODERN MATHEMATICS: THEIR NATURE AND USE, Jagjit Singh. Reader with only high school math will understand main mathematical ideas of modern physics, astronomy, genetics, psychology, evolution, etc., better than many who use them as tools, but comprehend little of their basic structure. Author uses his wide knowledge of non-mathematical fields in brilliant exposition of differential equations, matrices, group theory, logic, statistics, problems of mathematical foundations, imaginary numbers, vectors, etc. Original publication. 2 appendices. 2 indexes. 65 illustr. 322pp. 5⅜ x 8. S587 Paperbound **$1.75**

MATHEMATICS IN ACTION, O. G. Sutton. Everyone with a command of high school algebra will find this book one of the finest possible introductions to the application of mathematics to physical theory. Ballistics, numerical analysis, waves and wavelike phenomena, Fourier series, group concepts, fluid flow and aerodynamics, statistical measures, and meteorology are discussed with unusual clarity. Some calculus and differential equations theory is developed by the author for the reader's help in the more difficult sections. 88 figures. Index. viii + 236pp. 5⅜ x 8.
T440 Clothbound **$3.50**

THE FOURTH DIMENSION SIMPLY EXPLAINED, edited by H. P. Manning. 22 essays, originally Scientific American contest entries, that use a minimum of mathematics to explain aspects of 4-dimensional geometry: analogues to 3-dimensional space, 4-dimensional absurdities and curiosities (such as removing the contents of an egg without puncturing its shell), possible measurements and forms, etc. Introduction by the editor. Only book of its sort on a truly elementary level, excellent introduction to advanced works. 82 figures. 251pp. 5⅜ x 8.
T711 Paperbound **$1.35**

BOOKS EXPLAINING SCIENCE AND MATHEMATICS

Engineering, technology, applied science etc.

TEACH YOURSELF ELECTRICITY, C. W. Wilman. Electrical resistance, inductance, capacitance, magnets, chemical effects of current, alternating currents, generators and motors, transformers, rectifiers, much more. 230 questions, answers, worked examples. List of units. 115 illus. 194pp. 6⅞ x 4¼. Clothbound **$2.00**

ELEMENTARY METALLURGY AND METALLOGRAPHY, A. M. Shrager. Basic theory and descriptions of most of the fundamental manufacturing processes involved in metallurgy. Partial contents: the structure of metals; slip, plastic deformation, and recrystalization; iron ore and production of pig iron; chemistry involved in the metallurgy of iron and steel; basic processes such as the Bessemer treatment, open-hearth process, the electric arc furnace—with advantages and disadvantages of each; annealing, hardening, and tempering steel; copper, aluminum, magnesium, and their alloys. For freshman engineers, advanced students in technical high schools, etc. Index. Bibliography. 177 diagrams. 17 tables. 284 questions and problems. 27-page glossary. ix + 389pp. 5⅜ x 8. S138 Paperbound **$2.25**

BASIC ELECTRICITY, Prepared by the Bureau of Naval Personnel. Originally a training course text for U.S. Navy personnel, this book provides thorough coverage of the basic theory of electricity and its applications. Best book of its kind for either broad or more limited studies of electrical fundamentals . . . for classroom use or home study. Part 1 provides a more limited coverage of theory: fundamental concepts, batteries, the simple circuit, D.C. series and parallel circuits, conductors and wiring techniques, A.C. electricity, inductance and capacitance, etc. Part 2 applies theory to the structure of electrical machines—generators, motors, transformers, magnetic amplifiers. Also deals with more complicated instruments, synchros, servo-mechanisms. The concluding chapters cover electrical drawings and blueprints, wiring diagrams, technical manuals, and safety education. The book contains numerous questions for the student, with answers. Index and six appendices. 345 illustrations. x + 448pp. 6½ x 9¼. S973 Paperbound **$3.00**

BASIC ELECTRONICS, prepared by the U.S. Navy Training Publications Center. A thorough and comprehensive manual on the fundamentals of electronics. Written clearly, it is equally useful for self-study or course work for those with a knowledge of the principles of basic electricity. Partial contents: Operating Principles of the Electron Tube; Introduction to Transistors; Power Supplies for Electronic Equipment; Tuned Circuits; Electron-Tube Amplifiers; Audio Power Amplifiers; Oscillators; Transmitters; Transmission Lines; Antennas and Propagation; Introduction to Computers; and related topics. Appendix. Index. Hundreds of illustrations and diagrams. vi + 471pp. 6½ x 9¼. S1076 Paperbound **$2.75**

BASIC THEORY AND APPLICATION OF TRANSISTORS, Prepared by the U.S. Department of the Army. An introductory manual prepared for an army training program. One of the finest available surveys of theory and application of transistor design and operation. Minimal knowledge of physics and theory of electron tubes required. Suitable for textbook use, course supplement, or home study. Chapters: Introduction; fundamental theory of transistors; transistor amplifier fundamentals; parameters, equivalent circuits, and characteristic curves; bias stabilization; transistor analysis and comparison using characteristic curves and charts; audio amplifiers; tuned amplifiers; wide-band amplifiers; oscillators; pulse and switching circuits; modulation, mixing, and demodulation; and additional semiconductor devices. Unabridged, corrected edition. 240 schematic drawings, photographs, wiring diagrams, etc. 2 Appendices. Glossary. Index. 263pp. 6½ x 9¼. S380 Paperbound **$1.25**

TEACH YOURSELF HEAT ENGINES, E. De Ville. Measurement of heat, development of steam and internal combustion engines, efficiency of an engine, compression-ignition engines, production of steam, the ideal engine, much more. 318 exercises, answers, worked examples. Tables. 76 illus. 220pp. 6⅞ x 4¼. Clothbound **$2.00**

BOOKS EXPLAINING SCIENCE AND MATHEMATICS

Miscellaneous

ON THE SENSATIONS OF TONE, Hermann Helmholtz. This is an unmatched coordination of such fields as acoustical physics, physiology, experiment, history of music. It covers the entire gamut of musical tone. Partial contents: relation of musical science to acoustics, physical vs. physiological acoustics, composition of vibration, resonance, analysis of tones by sympathetic resonance, beats, chords, tonality, consonant chords, discords, progression of parts, etc. 33 appendixes discuss various aspects of sound, physics, acoustics, music, etc. Translated by A. J. Ellis. New introduction by Prof. Henry Margenau of Yale. 68 figures. 43 musical passages analyzed. Over 100 tables. Index. xix + 576pp. 6⅛ x 9¼. S114 Paperbound **$3.00**

MATHEMATICS, HISTORIES AND CLASSICS

HISTORY OF MATHEMATICS, D. E. Smith. Most comprehensive non-technical history of math in English. Discusses lives and works of over a thousand major and minor figures, with footnotes supplying technical information outside the book's scheme, and indicating disputed matters. Vol I: A chronological examination, from primitive concepts through Egypt, Babylonia, Greece, the Orient, Rome, the Middle Ages, the Renaissance, and up to 1900. Vol 2: The development of ideas in specific fields and problems, up through elementary calculus. Two volumes, total of 510 illustrations, 1355pp. 5⅜ x 8. Set boxed in attractive container. T429, 430 Paperbound, the set **$5.00**

A SHORT ACCOUNT OF THE HISTORY OF MATHEMATICS, W. W. R. Ball. Most readable non-technical history of mathematics treats lives, discoveries of every important figure from Egyptian, Phoenician mathematicians to late 19th century. Discusses schools of Ionia, Pythagoras, Athens, Cyzicus, Alexandria, Byzantium, systems of numeration; primitive arithmetic; Middle Ages, Renaissance, including Arabs, Bacon, Regiomontanus, Tartaglia, Cardan, Stevinus, Galileo, Kepler; modern mathematics of Descartes, Pascal, Wallis, Huygens, Newton, Leibnitz, d'Alembert, Euler, Lambert, Laplace, Legendre, Gauss, Hermite, Weierstrass, scores more. Index. 25 figures. 546pp. 5⅜ x 8. S630 Paperbound **$2.25**

A HISTORY OF GEOMETRICAL METHODS, J. L. Coolidge. Full, authoritative history of the techniques which men have employed in dealing with geometric questions . . . from ancient times to the modern development of projective geometry. Critical analyses of the original works. Contents: Synthetic Geometry—the early beginnings, Greek mathematics, non-Euclidean geometries, projective and descriptive geometry; Algebraic Geometry—extension of the system of linear coordinates, other systems of point coordinates, enumerative and birational geometry, etc.; and Differential Geometry—intrinsic geometry and moving axes, Gauss and the classical theory of surfaces, and projective and absolute differential geometry. The work of scores of geometers analyzed: Pythagoras, Archimedes, Newton, Descartes, Leibniz, Lobachevski, Riemann, Hilbert, Bernoulli, Schubert, Grassman, Klein, Cauchy, and many, many others. Extensive (24-page) bibliography. Index. 13 figures. xviii + 451pp. 5⅜ x 8½. S1006 Paperbound **$2.25**

THE MATHEMATICS OF GREAT AMATEURS, Julian Lowell Coolidge. Enlightening, often surprising, accounts of what can result from a non-professional preoccupation with mathematics. Chapters on Plato, Omar Khayyam and his work with cubic equations, Piero della Francesca, Albrecht Dürer, as the true discoverer of descriptive geometry, Leonardo da Vinci and his varied mathematical interests, John Napier, Baron of Merchiston, inventor of logarithms, Pascal, Diderot, l'Hospital, and seven others known primarily for contributions in other fields. Bibliography. 56 figures. viii + 211pp. 5⅜ x 8½. S1009 Paperbound **$1.50**

ART AND GEOMETRY, Wm. M. Ivins, Jr. A controversial study which propounds the view that the ideas of Greek philosophy and culture served not to stimulate, but to stifle the development of Western thought. Through an examination of Greek art and geometrical inquiries and Renaissance experiments, this book offers a concise history of the evolution of mathematical perspective and projective geometry. Discusses the work of Alberti, Dürer, Pelerin, Nicholas of Cusa, Kepler, Desargues, etc. in a wholly readable text of interest to the art historian, philosopher, mathematician, historian of science, and others. x + 113pp. 5⅜ x 8⅜. T941 Paperbound **$1.00**

A SOURCE BOOK IN MATHEMATICS, D. E. Smith. Great discoveries in math, from Renaissance to end of 19th century, in English translation. Read announcements by Dedekind, Gauss, Delamain, Pascal, Fermat, Newton, Abel, Lobachevsky, Bolyai, Riemann, De Moivre, Legendre, Laplace, others of discoveries about imaginary numbers, number congruence, slide rule, equations, symbolism, cubic algebraic equations, non-Euclidean forms of geometry, calculus, function theory, quaternions, etc. Succinct selections from 125 different treatises, articles, most unavailable elsewhere in English. Each article preceded by biographical, historical introduction. Vol. I: Fields of Number, Algebra. Index. 32 illus. 338pp. 5⅜ x 8. Vol. II: Fields of Geometry, Probability, Calculus, Functions, Quaternions. 83 illus. 432pp. 5⅜ x 8.
Vol. 1: S552 Paperbound **$2.00**
Vol. 2: S553 Paperbound **$2.00**
2 vol. set, **$4.00**

A COLLECTION OF MODERN MATHEMATICAL CLASSICS, edited by R. Bellman. 13 classic papers, complete in their original languages, by Hermite, Hardy and Littlewood, Tchebychef, Fejér, Fredholm, Fuchs, Hurwitz, Weyl, van der Pol, Birkhoff, Kellogg, von Neumann, and Hilbert. Each of these papers, collected here for the first time, triggered a burst of mathematical activity, providing useful new generalizations or stimulating fresh investigations. Topics discussed include classical analysis, periodic and almost periodic functions, analysis and number theory, integral equations, theory of approximation, non-linear differential equations, and functional analysis. Brief introductions and bibliographies to each paper. xii + 292pp. 6 x 9. S730 Paperbound **$2.00**

THE WORKS OF ARCHIMEDES, edited by **T. L. Heath.** All the known works of the great Greek mathematician are contained in this one volume, including the recently discovered Method of Archimedes. Contains: On Sphere & Cylinder, Measurement of a Circle, Spirals, Conoids, Spheroids, etc. This is the definitive edition of the greatest mathematical intellect of the ancient world. 186-page study by Heath discusses Archimedes and the history of Greek mathematics. Bibliography. 563pp. 5⅜ x 8. S9 Paperbound **$2.45**